# Cope with Your Biological Clock

# Cope with Your Biological Clock

## How to Make the Right Decision about Motherhood

Theresa Francis-Cheung

HELP YOURSELF

McGraw-Hill books are available at special quantity discounts to use as premiums and sales promotions, or for use in corporate training programs. For more information, please write to the Director of Special Sales, Professional Publishing, McGraw-Hill, Two Penn Plaza, New York, NY 10121-2298. Or contact your local bookstore.

0-07-139661-6

*Library of Congress Catalog Card Number*: On file.



First published in Great Britain in 2001
by Hodder Headline Ltd, 338 Euston Road, London, NW1 3BH.

This edition first published in 2002 by Contemporary Books,
a Division of The McGraw-Hill Companies.



Typeset by Avon Dataset Ltd, Bidford-on-Avon, Warks, England.

Printed and bound in Great Britain by
The Guernsey Press Co. Ltd, Channel Isles.

# Contents

# Acknowledgments

My thanks to all the women I talked to for the insight they gave me into biological-clock anxiety. Their stories are in this book, with names changed to protect their privacy. They provided much, if not all, the incentive to write about a subject which has for too long been dismissed.

I am indebted to all the staff at Hodder and Stoughton, in particular my editor, Judith Longman, for her charm, advice, encouragement and insight throughout. Thanks also to the fine developmental editing work of Dr Priscilla Stuckey and to my agent, Michael Alcock, for his feedback, advice, wit and friendship.

A big thank you for the support of family, friends, my brother Terry and his partner Robin. Finally, as always, special gratitude to my husband Ray, son Robert and daughter Ruth, for their patience, support, enthusiasm and love while I went into exile to complete this project.

# Introduction

*For a woman the pressure starts early. By 30, the family and in-laws are growing nervous: at 35 the ticking of your clock is threatening to wake the neighbours and by 40 you are considered past it and left in peace.*
(Deborah Holder, 'Primigravidas in their Prime', *Guardian* Women's Page, 23 January 1992)

Do you want to have a baby? Is it the right time to have a baby? Are you ready to have a baby? If so, why and when? Will you be a good mother? Can you afford to have a baby? If you decide not to have children, will you regret it? What if you're not interested in motherhood?

If you are in your thirties and have delayed making decisions about babies, these questions are likely to loom large. And if you are approaching forty you may be asking 'Can I have a baby? Have I left it too late?' The resulting anxiety can be so intense that it can distract you from other things in your life that matter, such as your job, your relationships and even your health. You may even fear that, under pressure, you'll decide to have a baby when you shouldn't, with someone who isn't right, or that you'll have to do it on your own.

This book is for you if you are contemplating motherhood and feeling anxious, uncertain and conflicted.

- You may have postponed making decisions about babies and feel pressure from your biological clock.
- You may want to have children, but feel that circumstances aren't right. For example, you may not have found the right partner, or developed your career sufficiently. You want to know how long you can safely postpone motherhood.
- You could feel ambivalent about children and want to see all sides of the picture before you make a decision.
- You could be contemplating single motherhood by choice.
- You may not be able to have children of your own. This book will help you understand that your fulfilment and happiness does not depend on becoming a mother.
- You may not want to have children. This book will show you that, contrary to society's expectations, this is a valid and healthy choice.

The aim of the book is to help you make the right decision about motherhood. It will explore the reasons you may have been delaying. It will help you discover if you are ready, willing and able to become a mother. It will show you that biological-clock anxiety is not something to be feared, but a necessary life crisis which can help you grow and develop into your full potential. The sense of urgency the clock brings forces you to make decisions about what you really want out of life – what will fulfil you as a woman.

Studies show that people cope better with stressful emotions and events when they feel adequately informed and supported. Hopefully by reading this book you will discover a sense of understanding, clarity and peace of mind that might have been missing before. You will learn that, even though you can't slow down or silence the ticking of the biological clock, you can deal positively with the stress while pondering your options.

I've written several books on aspects of women's health, but this one has gained more attention by far among my friends. One of them struggling with the baby-or-not dilemma told me that the book was a welcome relief. 'I was never really that keen on kids. I thought the maternal instinct would suddenly hit me. Well, I'm thirty-seven now and I'm still not sure. I have lived with such guilt.' Another told me, 'Thank goodness the subject is out in the open. I find it incredibly hard to admit that I really want a baby. I couldn't bear to miss out on the whole wonderful experience. But I don't have a boyfriend and am very aware of my biological clock ticking.' Yet another, who thinks she does want a baby, wondered if it was normal to have doubts. 'How will I know when I am ready?' One woman in her late twenties wondered if having her eggs frozen was the right thing to do. 'I'm worried that I will postpone and postpone until it's too late.'

I could go on and on with comments from women who find that the issue of baby or not causes guilt and anxiety, and who long to have an open, honest discussion. I am not talking here of dysfunctional women, of women who lack ambition, drive, discipline and challenge in their lives. I am not speaking of women who are consumed by their careers or who are obsessively preoccupied with themselves or with having babies. I am speaking, I believe, of the great majority of women, like you and me, who know their own minds and who are creating positive and exciting lives for themselves. I am speaking of women who want to find fulfilment, but who find that their greatest obstacle is not opportunity, but making decisions about babies.

The many expressions of relief I had when I discussed this subject convinced me I needed to write this book. I hope that as you read it you will feel better equipped to cope with the sometimes funny, sometimes stressful, sometimes terrifying and always complicated feelings that the ticking of your biological imperatives brings to the surface.

# 1

# That Ticking Sound

*I was still in that 'Babies!-God-not-yet' mindset. I assumed that everyone else was too. When my best friend told me that she was pregnant on purpose, I was rendered speechless. It really was like she had dropped a bomb on me. The sudden realisation that all over the world women my age had complete families of their own!*

(Belinda, age thirty-two)

You probably grew up expecting to have children one day. You didn't really think about motherhood that much. It was just something you thought would happen someday – after school, after college, after a year off, after establishing a career, after finding Mr Right, after making sure he is Mr Right. The someday of motherhood is always far off in the future.

There comes a point when you suddenly realise that 'someday' is now. That feeling of 'now' can strike at any time, in your twenties, thirties or forties. The feeling that one of life's biggest decisions needs to be made. It can be sparked by many things – by a partner, by a career break, by age and declining fertility, by a close friend who has a baby, by an unintended pregnancy, by a 'Baby Gap' ad.

At the moment of 'now' you may feel an overwhelming sense of shock if you have been lulled into a cosy sense of security of motherhood in the future. You may not be prepared for the flood of conflicting emotions you experience. You feel panic, alarm, fear, turmoil, anxiety as well as excitement, joy and anticipation. All of which can be summed up by the typical questions you are probably asking yourself, 'Am I ready?' 'Do I want this?' 'Can I do this?'

There are no right or wrong answers to the motherhood questions, but it is important you ask them. The decision to mother or not is one way for you to create your grown-up self. Whatever the ultimate decision, it is your journal of self-development. When you start asking yourself if you are ready to have a child, if you want a baby, if you can have a baby, you approach a major turning point in your life. You set the course for the rest of your life. The quest of this book is to help you deal with feelings of uncertainty, confusion and conflict during this time of crisis. The journey will require you to take an honest look at yourself, your relationship, if you are in one, and your lifestyle. Hopefully, by the time you reach the end of the book, you will feel better equipped to make a decision about motherhood that is right for you.

## Biological-clock anxiety

The phenomenon of biological-clock anxiety is a fairly new trend. A generation or so ago women tended to marry and have babies at a much younger age. But today we typically put off childbearing for a number of years in order to pursue further education or career or other goals. Then before we know it we are in our thirties or forties and anxious about our biological clocks. We may not even take the time to figure out if we actually want children or not. The mere fact that we may not be able to have a child, if we don't act soon, can completely cloud the issue.

Biological-clock anxiety won't necessarily make you irrational and chaotic, as we are so often led to believe, but the conflicting emotions the motherhood decision inspires will bring a real sense of crisis, challenge, opportunity and change to your life.

For some of us the rite of passage is a smooth one.

> 'I knew that I wanted to have a family by the time I was thirty-five,' says Virginia. 'That gave me time to enjoy my career and my marriage and earn enough money to afford a nice house. I didn't want to leave it to my late thirties because I was worried about my fertility declining. When I was thirty-four I stopped taking contraception and it took me six months to get pregnant. I took maternity leave for three months, and then had my second child a year later. I'm working four days a week now and am comfortable with my day-care arrangements. I'm happy with how everything is working out.'

> In contrast, there is Lucy who decided in her late twenties that she wasn't going to have any children. She's forty-seven now and hasn't regretted her decision. 'I just knew that I didn't want to be a mother. My job is demanding and the hours long. I love to travel and I love my freedom. I don't have enough room in my life for children.'

For the majority of us, though, life simply isn't that straightforward; things just don't fall into place so smoothly.

> Janice decided she wanted to start a family when she was thirty-two, so that she could be back at work full time by the time she was thirty-five. She decided to stop taking the pill. Two years later she still hasn't conceived and has had one miscarriage. 'I thought I could plan everything, but life doesn't work that way. I thought I had the choice, but nature has a way of choosing for you sometimes.'

But even if you do feel you have a choice, making that choice isn't easy. You are probably intelligent enough to realise that just wanting to have a baby is not enough. Wanting a baby is not the same as being emotionally, physically and mentally prepared to have one. Nor does the desire to become a mother mean that you will be a good one. You may be plagued by doubts and fears.

> 'I always thought I'd have children,' says Victoria, age thirty-six. 'I know I need to make a decision soon, but I really don't know what to do. I'm not sure if I will be a good mum. I don't know if I want to give up my freedom. I'm not even sure I'm in the right relationship.'

You may be keenly aware that some women make better mothers than others. With the stresses of modern living you could be uncertain whether you have the money, time, strength, patience and selflessness to raise children. You may even wonder if you really want to bring children into this chaotic, stressful, overcrowded world.

On the other hand, you may not be sure if motherhood is for you at all. As time ticks away on your biological clock you may wonder why children are not a priority in your life. Why you don't feel a sense of urgency. Isn't it the natural thing for a woman to want? You may worry that there is something wrong with you.

Or you may know that you want babies, but the circumstances or the relationship isn't right, or you haven't found the 'right' man. Your anxiety centres around how long you can reasonably postpone. You may be delaying decisions about childbearing through fear – fear of childbirth, fear of loss of identity, fear of weight gain, fear of the loss of free time, fear of loss of intimacy in your relationship, fear of missing out career-wise. Or you could have problems conceiving.

You could be a mother already struggling with the decision to have another child.

> Amanda had a child from her first marriage. He is a teenager now and very independent. She is thirty-nine and recently got remarried. 'I really want to have another baby to seal the relationship. This is really important to us, especially because James hasn't got any children himself. I'm worried that I might have left it too late.'

> Janet is thirty-five. She had her first baby three years ago and

> really wants to have another. But with the stresses of her job, the problems with childcare arrangements, and her husband's low income she knows she can't really afford another child.

Or you could be the kind of woman who only feels fulfilled if you have a small baby at your breast.

> 'I had three children in my late twenties and early thirties. I loved the experience so much I just had to do it again and again before it was too late,' says Marilyn, mother of five. 'I'm thirty-seven now and, if my husband agreed, would have a sixth child. I do wonder if I am wanting one just for my own pleasure, but am I really doing something so very wrong?'

Whatever your circumstances, the older you get the louder your biological clock will begin to tick. It will remind you constantly that decisions about motherhood need to be made – and they need to be made soon.

## Men and biological-clock anxiety

'Single, thirtysomething, broody – yes that's him' (Shane Wilson, *The Evening Standard*, 28 February 2000, p. 29) – you may think that men don't have to worry about their biological clocks, but anxiety at the prospect of dwindling childbearing years is not confined to the female species.

We have seen a boon in the 'new man' idea in films, TV and video. When celebrities like David Beckham take time off from their sporting schedule to nurse an ill child the media goes into a frenzy. It is seen as a symbol of changing times. There is a long way to go, but more men are taking their roles as daddies seriously and actively involving themselves in childcare.

And along with these devoted new fathers we now have the single, broody man. The easygoing man who thinks he has years to settle down is still in the majority, but it seems the biological clock is starting to tick louder for some men than women. Bridget Jones's search for Mr Right and babies is being replaced by the single man's longing to enhance his life with children. Some men are starting to admit that they not only want babies, but they want to love and nurture them too.

> Martin, age thirty-five, says, 'I want children soon. Really because it feels like something to do at my age. I want to be a dad like everyone else. I think having children is what makes you a man. The trouble is my girlfriend doesn't want to take a career break.'

Then there is the older man with the younger woman who fears being an 'old dad'. With a partner nonchalant about procreating, he spends his time in Baby Gap dreaming of taking his child to that first football match. I spoke to several women who had babies a few years earlier than in their game plan because their partners were several years older.

Women are becoming warier about motherhood. It's fine for a working man to say that his children are everything to him, but it's still not acceptable for working women to make children their priority. Having children tends to change us in the world's eyes. There can be a loss of status. But it doesn't change men or their status. For a man children can be something to look forward to and enjoy. It's not quite so simple for us. Our whole lives are turned inside out.

If the trend to postpone or not have children continues, male biological-clock anxiety might become more commonplace. Who knows, perhaps an increasing number of broody men desperately trying to convince women to have their babies will be the trigger society needs to start rewarding and valuing motherhood more.

For the present, however, it's unlikely that things will head in that direction. Biological-clock anxiety is still far more likely to be a source of stress for women. Yes, men have ticking clocks too, but at the end of the day, when we start hearing that ticking sound, we know our options will run out sooner than they will for a man. The pressure really is on us to decide, before menopause makes up our minds for us.

But what if you can't make up your mind?

## A lonely decision

The question of children or not is so inextricably bound to the idea of fulfilment as women that it can create incredible tension if you are faced with a limited amount of time to make up your mind. Because motherhood is so often considered a natural choice, and you are expected to 'just know', it can be incredibly stressful when the feeling of 'now' strikes, and you don't know what to do. And, unfortunately, when you reach this crucial, anxious moment in your life there is often no one to confide in. You may feel, as I did, very lonely.

> In my twenties I was dead set against children. I wanted to be a dancer. I knew that my career would be demanding and short. Babies would get in the way. They would ruin my body. Maybe sometime in the distant, distant future. But certainly not now.
>
> In my late twenties I did occasionally find myself looking curiously at babies in prams. No longer a dancer, but a writer, teacher and journalist, I was still in the pregnancy-is-something-to-be-avoided-at-all-costs mindset. I began to think at times that I just wasn't cut out for motherhood. I didn't tell anyone how uncomfortable children made me feel. They were noisy, demanding and chaotic. I thought there must be something wrong with me.
>
> The first big shock came when a colleague of mine got married and had her first baby. I couldn't believe it. I remember thinking, 'But she's only twenty-nine', and feeling faintly ridiculous. Getting pregnant wasn't really that big a deal when you were pushing thirty!
>
> I started to hear about more and more women my age getting pregnant. Friends, colleagues, my cousin, even my hairdresser.

Not only were they getting pregnant intentionally, but to make matters worse they were thrilled at the prospect. They talked incessantly about the baby growing inside them. How giving up smoking, drinking and late nights didn't matter. The important thing was to have a healthy baby.

The notion of actually making the decision to have a child filled me with terror. Grown-ups got married and had children and I didn't feel that grown-up. It seemed that time had moved on without me. Although I still found the whole baby business a bore, I knew deep down that I hadn't dismissed the idea of children completely. What I had done was repress the idea of reproduction, convinced that I wasn't grown-up enough. But now friends in my age group, with similar lifestyles to mine, were having babies! They assumed that one day I would have them too.

For the first time in my life my hectic, busy schedule began to lose some of its appeal. I would get tearful if I came home to an empty house after a long day and there were no messages on the answer phone. I was in my thirties and my biological clock was ticking. What did I want? I really didn't know. I couldn't see myself as a mother, but I also couldn't see myself not being a mother. The whole thing was just too difficult and heated a subject to discuss with anyone. I kept my anxieties to myself.

A year or so later I got married. It wasn't long before we started trying for a baby. I had my doubts about becoming a mother, but I didn't talk about them, not even to my husband.

Two years ago, after the birth of my son, if you had told me that I would be writing a book about biological-clock anxiety I would have found the notion intriguing. Surely that stage of my life was in the past? I was perfectly content with my choice to become a mother. But last year when one of my oldest friends casually remarked that she and her partner were thinking about having a baby, and wondering if it was the right thing to do, I wasn't prepared for the range of emotions I began to experience.

- **Concern**: I didn't want my friend to be burdened with the responsibility of parenthood. I didn't want her to feel the pain, the guilt and the anxiety.
- **Fear**: what if she couldn't have children? How would that make her feel? How would it affect our friendship?
- **Relief**: now I wouldn't have to envy her freedom any more. Now she would understand what I was going through – how stressful and stifling motherhood can be.
- **Jealousy**: she was going to experience the incredible and wonderful emotions of pregnancy and motherhood.
- **Anxiety**: I wasn't getting any younger. Did I want another child? Could I cope with another child? Would my friend be able to cope with motherhood?
- **Selfishness**: I hope she wasn't going to depend on me too much for advice, baby-sitting etc. I just don't have the time or energy.
- **Sorrow**: my friend had no idea of what she would be giving up, how her life would change.
- **Joy**: I was so happy for her. She would know the joy, wonder and fulfilment of motherhood.
- **Insecurity**: perhaps she would cope with motherhood better than me?
- **Worry**: did my friend know how expensive having kids is? Did she know how dangerous it was to smoke during pregnancy? How important it was for her to take folic acid? How her relationship with her partner would change? How her attitude to work would change? How her figure would change? How her life would change?

I wanted to tell her so many things. Things that had been on my mind ever since I started hearing my own biological clock ticking. I have never regretted the choices I made, but I wish that I had been more prepared, that I hadn't been so alone with my worries. Yet I found it incredibly hard to give my friend any insight at all. All I could come up with were typically bland and inadequate clichés like, 'How wonderful!' 'You'll love being a mum.'

It is rare for mothers to warn inexperienced friends or relatives about how much children will change your life. How having a baby can, but does not necessarily, offer complete fulfilment. It is equally rare for child-free women to talk frankly about how their decision not to have children is shaping or has shaped their lives. How being child-free can, but does not necessarily, offer complete satisfaction.

As a result, when you start to hear your biological clock tick you'll get little advice and insight and lots of clichés and half-truths. Yes, for some women motherhood fulfils all expectations. Yes, for some women a child-free lifestyle is the happiest choice. But there are also many mothers who feel overwhelmed and many childless women with regrets.

Those of you yearning for a child who read this book may wonder what it is I am talking about. Women who don't think they want kids may also wonder. But what I am talking about is the truth: that there are both painful and wonderful aspects to motherhood and to the child-free lifestyle. I want women who hear their biological clock ticking, and aren't sure about motherhood, to see both sides of the coin, to be a little better informed before they make one of the most important decisions of their life.

Chapter 4 will show you that whatever choice you eventually make there will be both losses and gains. But first, it will be helpful to explore some of the reasons why, when biological-clock anxiety strikes, you may feel, as I did, so terribly vulnerable and alone. Why don't we like talking to each other about the decision to mother or not? Why is it we don't help each other when our need for guidance and consolation is so great?

# 2

# Why We Don't Like Talking about the Clock

*Wish Jude would not talk about biological clock in public. Obviously one worries about such things in private and tries to pretend whole undignified situation isn't happening. Bringing it up in '192' makes one panic and feel like a walking cliché.*

(Helen Fielding, *Bridget Jones: The Edge of Reason* (Picador, London), p. 39)

You probably don't feel comfortable talking about your biological clock. You find it hard to admit that decisions about motherhood are causing you anxiety. Why? Here are some possible reasons.

## 'Just you wait till you have children'

'I do get depressed sometimes,' says Susan, age thirty-three. 'Maybe kids are the answer, but maybe they aren't. There's no point talking about it though. I'm at that age, you see, when it is assumed that all you really want to do is have kids and bake cakes.'

Motherhood is no longer the 'norm' for women today. But even though you have so many life choices available to you, the myth of maternal instinct and the rewards of motherhood have been so hugely magnified that, however fulfilling your job, your interests, or your social life, however ambivalent you are about having children, there is always a little voice telling you that you really ought to have kids.

> Rebecca is thirty-six years old. She recently got married and moved into her partner's flat. When she returned from the honeymoon a trunk sent by her grandmother had arrived. It was full of baby clothes and toys. 'It made me feel very anxious,' she says. 'How could I break the news to my family that we hadn't made up our minds about having children?'

Every one of us hears the real or imagined voice of mum or grand-mum urging us to fulfil our role as a woman: to become a mother. Society expects us to become mothers. We get so used to living with the pressure that we may not even notice it any more. But just like the air we breathe, this expectation invisibly feeds our lives.

The social pressure to mother, the assumption that you really want to 'settle down' and have children, can sometimes be over-whelming. It certainly inhibits frank discussion about the decision to mother. 'When you have children of your own' is a familiar phrase that reinforces the inevitability of motherhood. If you babysit for a friend you are 'practising' for the time when you will mother. 'Do you have children?' is often asked at social functions.

There is an immense weight of tradition that assumes that any normal woman is born with a maternal instinct – that this maternal instinct is what it means to be born female. Normal women are supposed to experience an urge to mother, to protect and to nurture children. If you don't have children you may find yourself accused of not fulfilling your natural function, as if pregnancy is in your best interests.

Throughout your life the maternal instinct is exalted by myths, folklore, custom, religion, social institutions, art, and by family and friends. You probably grew up playing with dolls because girls 'like that kind of thing'. When you go to work the assumption is usually that you will give up work once you have children. Marriage is still regarded as a prerequisite for you to have babies. Casual remarks, should you get married, reinforce the inevitability of children. 'It'll be your turn next.' If you are single you'll probably feel direct pressure from family and friends to settle down.

Every time a celebrity is interviewed by a women's magazine, the question of whether or not she will have children is brought up. You are led to believe this is your function in life. Feminism has done much to expose the hardships of motherhood, but this has not quenched the media's thirst for idealising the joy and fulfilment of motherhood and family life. It is hard for you to have an objective image when you are totally surrounded by it.

Images of women who can't, or choose not to, have children are seldom positive. There is the sad, unfulfilled, barren woman full of regret. Or the neurotic, single-minded bitch who is selfish, unnatural and eccentric. This is your fate if you decide not to fulfil your biological destiny. Even if the image presented is slightly more positive, an underlying unhappiness and insecurity is usually hinted at.

At the root of all this pressure is the social 'norm' about women which has become internalised as part of your psychological make-up. Normal women are supposed to want to mother. It is only through maternity that you can find fulfilment.

> By age 30, women who earlier had no special desire for motherhood often start having maternal wishes. According to a popular saying 'She hears the biological clock ticking.' (Daniel Levison, *The Seasons of a Woman's Life* (Alfred A. Knopf, New York, 1996), p. 365)

But what exactly is this maternal urge? Is it biological? There is no doubt your body is designed to reproduce. The future of our species depends on it. Could it be then that the maternal instinct is an emotional, instinctual, unconscious, hormonal or tribal urge you are powerless to resist?

Many women who become mothers say that it made them feel

normal and feminine. That it completed their identity as women. That they joined the sisterhood of mothers. Without a doubt nurturing babies is fundamental to their fulfilment. For women who have identified with the role of mother, the desire for a baby can feel so instinctive that it feels inborn, but to suggest that because you can mother you will be fulfilled by motherhood is a ridiculous notion.

Theories of biological determinism have trapped women for centuries. The argument that the maternal instinct is governed by culturally determined factors has been one of the great crusades of feminism. In the last fifty years or so the myth about women being fulfilled by motherhood has been exploded by writers like Simone de Beauvoir, Betty Friedan, Germaine Greer and Shere Hite.

One of the major arguments against the existence of mother instinct is, if such an instinct exists, why have so many women chosen not to have large families, or even to start families, when contraception became available? And if motherhood is so rewarding, why are so many women with children depressed? And what's more, now that we have the power to choose, denied to our grandparents, is mothering still a natural part of a woman's life?

We assume that the mothering instinct is genetic or hormonal, but this isn't necessarily the case. Much of it is influenced by experience and what we know. Perhaps we need to look at what it is in a woman's life, what it is about her growing up, that sways her for or against children. Each woman's attitude towards motherhood will be shaped by her individual life experiences. It is becoming apparent, as more and more of us choose not to have children, that we are not all born with an urge to mother that suddenly appears and takes over our lives. What we are born with is the ability to reproduce should we choose to. And that ability to reproduce is not necessarily essential to our fulfilment.

But the indoctrination is so great that many of the women I talked to during the writing of this book believed that the maternal instinct would suddenly develop and override all their anxieties about motherhood. They believed they would be struck by an uncontrollable desire to have children – the feeling that maybe not now, but one day, they would start wanting to have a child. Most thought that by the time they reached their forties they would get broody. Those ambivalent about children admitted that they felt pressured to have children, because they feared that broodiness or a sense of regret would strike in later years.

> 'When a woman starts feeling broody she wants to become a mother. She stares longingly at babies in prams and fantasises about being pregnant and starting a family. The need to mother becomes a priority in her life. This may be a passing phase or it may linger for years.'

There certainly are women who do experience intense longings for babies, but the idea that every woman falls victim to bouts of broodiness is as much a myth as the myth of the maternal instinct.

> 'I thought there was something wrong with me,' says Sheila. 'I kept waiting for the maternal instinct to strike – for me to go all weepy whenever I saw a baby. Well, I'm forty-six now and I can honestly say that it never did strike. I'm not comfortable around babies and children.'

Feeling broody can mean that you are ready to become a mother, but then again it may not. Because our culture places so much emphasis on motherhood as being central to a woman's life, it can be hard to know exactly what it is you really want when you start thinking about having babies.

Is the desire to mother really a desire for babies, or the result of social pressure? The desire to fit in? Is it the urge to create? Is it the urge to be pregnant? Is it the urge to mother? Is it the urge to be like other women? Is it the urge to prove that you have the body of a normal woman? Is it the need for a family? Is it the need to love and be loved? Is it the need to feel whole? Is it boredom or lack of fulfilment? Is it the need to fill a void in your life?

This is not to say that there are always complicated undertones when you start thinking about children. It is simply to say that having children may not always be the ideal remedy for your feelings of restlessness. Thinking about babies and the possibility of motherhood may just mean that you have reached a stage in your life when you long for greater fulfilment. You may choose to find this fulfilment through motherhood, but there may also be other options you should explore.

Coping with biological-clock anxiety is all about you asking yourself what your feelings mean to you. Not what is considered normal by others, but what is right for you. It involves understanding that any definition of fulfilment, of being a whole or a normal woman, that defines itself by motherhood is misleading.

Coming to this understanding, however, and having an honest discussion, free of prejudice and assumption, is not easy in a society that repeatedly tells you 'You will be mother.'

## A walking cliché

> 'I spend a lot of time thinking how much I want to have children. I feel lonely and left out, because that's what my age group seems to be doing. It's hard to meet with women who will admit to feeling this way too. They don't want to look desperate.' (Debbie, age thirty-four)

Another reason you may find it hard to talk about biological-clock anxiety is fear of looking vulnerable, needy and neurotic, of becoming a walking cliché. The notion of the childless woman over thirty as a biological time-bomb with a dicky fuse is deeply ingrained.

When I hit that stage in my life, all I could think about was Glenn Close in *Fatal Attraction* boiling the pet rabbit of her married lover's son. The film hints at the character's infertility.

Or the sad, sadistic nanny, barren after a miscarriage, played by Rebecca de Mornay in *The Hand that Rocks the Cradle*. These images burned into my brain when I first saw them in my early twenties. Try as I might I couldn't erase them.

There is a moment in the popular film *When Harry Met Sally* when Meg Ryan discusses relationship problems with her friend.

'You're thirty-one. The clock is ticking,' warns her friend. 'No, it isn't,' Sally replies, looking uncertain and terrified. 'I read it doesn't start until you're thirty-six.'

Ally McBeal is forever paranoid about her biological clock and in the hugely successful *Bridget Jones's Diary* (1997) the reader is keenly aware that time is running out for the insecure, anxious heroine. Matt Groening's female rabbit standing on a street corner with a sign that reads 'I am 30 now and I want to have children' in his brilliant comic strip series *Life in Hell* sums up the general tone.

It isn't really cool to talk about your ticking biological clock. It's embarrassing and it makes other people feel uncomfortable. It's something you prefer to worry about in private. It's something everyone else, including family and friends, would prefer you worried about in private too.

## A woman thing

> 'I really like Paul,' says Susan, age thirty-six. 'I think he is keen on me too. I want to play it cool though. You see, he's five years younger and I'm worried that he might think I'm desperate to have kids. He may not be ready for that yet. I don't want to scare him off.'

If talking to family and friends about biological-clock anxiety is hard, talking to the men in your life is even harder.

Later in this book we will discuss the single motherhood option. On the whole, though, for many of us finding the right man is often the determining factor. You probably want your baby to have a father. But actually bringing up the subject of children is another matter!

You may want children, but can't talk about it to your partner for fear of losing him. (There are exceptions, but most men don't seem to be in the same kind of hurry to commit and have children as we are.) You may be with a partner who doesn't want children and this is causing conflict. Or you may not be convinced that your partner is father material. Then again, you may not have a partner at all.

And even if you think you have the right man, you don't win there either. Because of the way our societies and economies are structured, more of the work of raising children falls to us. It is our lives that will change the most.

When I got married we talked about children and I knew that my husband was vaguely committed to the idea. But at the end of the day I knew that, although his life would change and he would also be responsible for raising our children, it was my body and my life that were going to be impacted in ways that he would never be able to relate to. Ultimately *I* would have to decide whether or not I was prepared to transform my life in such a way.

## Mothers v non-mothers

> 'My friends with kids just don't have time for me any more. I do understand that it is hard for them to socialise like they used to, but I also feel rejected. They obviously feel that I don't really understand the momentous thing that has happened to them. We both feel let down.' (Lara, age thirty)

You would think that you could turn to friends who have made the decision to mother or not for support, insight and advice, but sadly this doesn't usually happen.

Mothers are often scared to admit that they may have feelings of ambivalence and resentment towards their children. On the other hand, they may also have problems admitting to you the unprecedented depth of love, sense of wonder and renewed faith in humanity that a child can bring into their life. Why else would a woman lose so much sleep? Why else would she let a newborn invade her every waking moment? Why else would she respond so automatically and rapidly to her newborn's needs? Fear that partners might become unbearably jealous could be a contributory factor, as could a woman's concern that she might seem tactless mentioning it to women yearning for children. But more likely it is our limited vocabulary of love that inhibits discussion. It is acceptable for a woman to fall in love with her partner, but talk of falling in love with your baby, of losing yourself in your child, might be considered unliberated in this age of individual achievement for women.

Women who have decided not to have children may feel reluctant to tell you that, at times, the child-free choice can make them feel vulnerable, lonely and exposed. On the other hand, for fear of offending, they may not tell you how incredibly relieved they often feel to escape the stresses of pregnancy, childbirth and motherhood. So ingrained are social expectations about motherhood that talk of liberation and having an identity outside the family might be considered tactless, selfish, unnatural and eccentric.

Having a baby changes your life. Babies are time-consuming and demanding. They tend to become the main focus and pattern of one's existence. We always gravitate towards the social group to which we have most in common, and mothers find great support from relationships with each other. If you haven't decided about children yet, you may find yourself dropped from the social calendar.

It works both ways though. You may find that you are withdrawing from friends who become mothers. You may become bored with the baby-oriented conversation and find the noisy demands of children unbearable. Or if you want children and circumstances don't allow, you may find it hard to deal with your feelings of regret.

How motherhood affects friendships depends on the individuals concerned. From my interviews with mothers and child-free women I often detected envy and resentment on both sides. Many mothers resented the loss of the ease and freedom of their life. Many child-free women felt they were missing out.

> 'My supervisor took me out to lunch to discuss a new project. It was the first time I had met her. I'm a freelancer and work from home so I can look after my kids. We talked about her busy schedule, her active social life and the fun she was having. I felt more than a twinge of envy. My life seemed so dull. But when the conversation turned to my children the tables turned. She actually told me how much she envied me. There we were, the two of us, both wanting what the other had.' (Mandy, age thirty-nine)

Some women, realising the possible differences that might occur when friends have babies, make a great effort to sustain the relationship and the rewards are great. But for all too many the potential tension means that it is easier for mothers and child-free women not to continue or initiate friendships.

> Ruth, a 36-year-old teacher, finds that she deliberately seeks out friends who don't have children. She finds that they are easier to socialise with and more emotionally available to her.

> Laura, a 37-year-old mother of two, finds it so much easier to break the ice if a woman has children too. She does have a few friendships with child-free women, but admits that they are not really close.

Despite motherhood's ubiquity and the ease with which it is entered into, it remains the great legitimiser among women. When I became pregnant I found that it was almost like joining a church – where the members would not just welcome you but embrace you as if you had seen the light and chosen the better path. My friendships when I didn't have children had much the same kind of self-affirmation.

Mothers tend to have a conviction that they are more womanly, more natural, more fulfilled than women who are not mothers. Child-free women tend to have a conviction that they are truly liberated, more fulfilled, more in charge of their destiny than women who are mothers. The smugness of both these positions offends me, but what offends me more is their flip sides. Women without kids are nothing. Women with kids have no identity. Whatever happened to the idea that women should no longer be judged by the use they make of their reproductive organs?

Few of us want to be outsiders. We all like to feel that we belong to a group of people and that we are accepted. That the choices we have made are the right ones. The problem with this perfectly understandable defence mechanism is that it creates a divide between mothers and non-mothers. It puts up barriers which inhibit honest and open discussion. It prevents the undecided among us from gaining real insight about the life choices other women make.

## The masks we wear

Have you noticed how delighted some mothers are when they hear that a friend is having her first baby? Some of this delight may be genuine, but some of it may be relief that they don't have to envy her freedom any more.

Feelings of misogyny are as deeply embedded in the female psyche as the male, perhaps, explains feminist theorist Dorothy Dinnerstein, because when we were small we were all helplessly dependent on an infinitely powerful mother. Or perhaps social pressure to conform, to appear 'normal', is so strong that we are terrified to speak our minds. Maybe we can't admit to each other that, despite having so much choice in our lives, a part of us still longs to do what women have always done.

> 'I didn't dare tell anyone about it, especially not my determinedly career-minded friends. We were educated, ambitious, intelligent women. We had worked hard to make something of ourselves. Even those who were married would have recoiled in horror and shuddered at the thought of kids. To admit that I was struggling with the baby-or-not dilemma seemed so pathetic. Such an admission of failure.' (Jo, age thirty-three)

Whatever the reason, a mutual fear and distrust often exists between women. At times of extreme vulnerability we prefer to keep silent about the truth of our lives. We are seldom truly honest and open with each other about the life choices we make.

Thankfully there has been a trend recently to be more truthful about our lives. Books, articles, films and TV specials about the horrors of raising a child, or the anxieties single, childless women face, take the lead. The truth is often revealed through humour. We have Bridget Jones and Ally McBeal with their biological-clock angst. Witty, truthful portraits of pregnancy and the first year of motherhood, like Vicki Iovine's *Girlfriend's Guides*. Frank accounts of the boredom and frustration of motherhood, like those by Kate Figes and the late Erma Bombeck. Susan Jeffers' *I'm Okay . . . You're a Brat* is a wonderfully cynical, refreshing guide for parents born without what Jeffers calls the 'Loving-Being-a-Parent' (LBP) gene.

But it will take much more than a few books, articles and films for attitudes to change. Progress can only really be made when each one of us starts to be more truthful about our own lives. Not being totally certain about every aspect of your life can make you feel uneasy. But if you don't present the full picture the result is confusion. If you just accentuate the positive and don't talk about the negative you don't do yourself, or the people you love, any favours.

Every life choice is confusing. There is no such thing as absolute certainty in life. You can't always be certain about what you really want. You can want a baby, but feel daunted by the responsibilities of parenthood. You can love being single and child-free, but still have your regrets.

As far as the parenthood decision is concerned, admitting that you have conflicting feelings and appreciating that there are two sides of the coin will help you make wiser choices. It will help you avoid the confusion of unrealistic expectations and choose motherhood or not with open eyes.

Now that we have shed some light on why women don't like talking about the clock, and how honest, open discussion could ease some of the anxiety, let's explore the many undefined and often unexpressed reasons women decide to have children, don't have children or aren't sure they want children. You may be shocked to discover that some of these reasons given are not as simple, natural or as wholesome as we would have one another believe.

If you are experiencing conflicting emotions about why you both want and do not want babies the next chapter will show you that you are not alone. The motivations behind the decision to mother or not rarely, if ever, have a black and white feel to them.

# 3

# Brooding on Breeding

*It's tough isn't it? Teasing out the different strands of thought and feeling, trying to get in touch with what we really want?*

(Jane Bartlett, *Will You Be Mother?* (Virago, London, 1994), p. 103)

How do you decide if you want children? How do you decide that you don't want children?

Clearly the paths, both emotional and social, that lead to the baby-or-not decision for every individual woman, are many and complex, reflecting the importance of what motherhood means to a woman. There is never one single reason for a woman to choose to mother or not, but a number of reasons based on her unique life history. So bear in mind that what follows, although intended to give you insight, is nothing more than a snapshot of the flood of undefined feelings, reasons and motivations that overwhelm women when they realise a choice has to be made.

## Making the decision to mother

> 'I don't know, it's hard to describe,' replies Janet, age thirty-six. 'I just knew the time was right. I felt ready.'

Many of the mothers I spoke to mentioned a feeling of 'readiness', of just knowing, as Janet did, that they wanted to have a baby. When I asked them what made them know they were ready, the responses were varied.

Having a loving partner that the option has been discussed with, and feeling that the child would be a natural extension of mutual love and respect, was a decisive factor for some.

Other mothers said that their decision to mother had been influenced by financial security. Being able to afford a baby and give it the best that they possibly could was important to them. For many, however, the optimum motivation to have a child was knowing that they had the time, energy and, above all, maturity to take on the responsibilities of motherhood. They felt they knew all about the positives and negatives of motherhood. They had a good deal of insight about what they were getting into. They knew that this was something that they and not their family, friends or even their partners wanted.

> Carol, age thirty-four, stressed the importance of feeling emotionally ready, knowing that she was able to put the needs of her child above her own and that she was mature enough to do whatever she needed to give her child love and support without resentment.
>
> 'I had healed a lot of old wounds from the past. I knew enough about myself to know that I had it in me to be a good mother. I knew that I could put the needs of another above my own because I was content and happy with my life. I did not expect a child to give it instant meaning.'

Many mothers admitted that, although their lives would be easier without children, they had reached a point when they were tired of thinking about themselves all the time. They were bored with endless self-gratification. They wanted to think about someone else for a change. They felt that their lives would be enriched by a child. Although they would have to put someone else first, the sacrifice would be paid back a hundredfold.

Others thought it was selfish not to have children. That the human race would cease to exist. That children were creating and funding our future.

Other women said that they had always wanted children and simply wanted to pass on their love and wisdom to them. They loved children and could think of nothing more satisfying in life than watching them grow and develop into well-balanced adults.

> 'Whatever you do in life,' says Sasha, age twenty-nine, 'nothing gives it more meaning than parenting. At the end of the day, whatever my achievements, the most important thing will be to have "loving mother" on my tombstone.'

Clearly for many women much thought and deliberation went into the motherhood decision. For others, however, this didn't seem to be the case. The decision to have babies was automatic. You get married. You have babies. There was no conscious decision. Socialisation continues to reinforce the expectation that most women want babies. It is simply something women do. 'I couldn't imagine not having children' was a persistent theme.

Many women who wanted, or had children already, gave answers compatible with the conditioning we receive to become nurturers. 'I want a family.' 'I adore children.' 'I want to love someone.' Others gave answers that contradict the notion of mother as nurturer. Several suggested that they were the ones who wanted to be loved.

> 'I went from disastrous affair to disastrous affair,' says Sally, age thirty-two. 'I was looking for love, but didn't find it. I started to think that if I had a baby I would have someone who loved, needed and wanted me.'

A few wanted to feel young again, to relive their childhood in a happier, more idealistic way or to undo negative aspects of their childhood. Others said they hoped a baby would bring them fulfilment, giving a sense of purpose and meaning to their life.

Some women felt that a career can only be purchased at the expense of love, security and motherhood. They couldn't envisage the possibility that they could be successful at both. Faced with a choice, they chose motherhood.

'Feeling like a real woman' was an often repeated phrase when it came to the baby decision, as if having a baby somehow made a woman more feminine. For the older mother, getting pregnant in her late thirties or early forties was certainly a source of pride. Gail Sheehy suggests in *New Passages* (HarperCollins, London, 1996) that in women of a certain age a new form of one-up-womanship is developing: Who has the youngest child? Who is still young and fertile?

Having a baby to fill a void in one's life, to provide a sense of purpose, to escape from the pressures of work or to prove something to others are questionable motives. They place an unfair burden on the child. Equally questionable is another reason sometimes given for having a child: the hope that it will bring a couple closer together. A baby can give a couple common purpose, but if a relationship is doubtful to start with, far from cementing the relationship, a baby just adds to the stress. Frequently it drives couples further apart, especially if the man feels that the woman got pregnant to 'trap' him.

Pressure from the biological clock, when a woman hits her thirties or forties, is clearly a motivating factor for some women. They think they want children, but are not sure that now is the right time for a variety of personal reasons, be it career pressure, finance, problems in relationships and so on. But at the same time they feel they can't afford to wait.

> 'I didn't start to panic until I got to thirty-eight,' says Sarah. 'I knew I wanted kids. I didn't feel ready, but I decided that I couldn't risk waiting any longer. If it wasn't for my

> biological clock ticking I probably would have procrastinated a while longer.'

Biological-clock anxiety puts some women in panic mode, which is not conducive to making a wise choice. Having delayed the baby decision until their late thirties, there comes a furious scrambling to make up for lost time. Fear of infertility and approaching menopause can cloud judgment.

> Marsha, age forty-one, admitted that she had been so focused on being able to have a child that she never really asked herself if she actually wanted a child. 'It never crossed my mind to think that maybe there had been a reason why I had delayed having babies. I was so obsessed with my fertility declining that was all I could think about.'

Other women said they felt pressured when partners or family urged them to have children or when friends started having babies.

> 'I don't think I really ever gave it any conscious thought. My mum wanted to be a grandmum before she was fifty and my husband wanted children. How could I disappoint the people I loved and respected the most?' (Linda, age thirty)

> 'The trigger for me to have kids', says Lucy, age thirty-five, 'was that everyone else seemed to be having them. All of a sudden I seemed to be the only one not to be married and having kids. I remember how jealous I was when my best friend had twins. I didn't want to feel left out.'

The media's obsession with motherhood was also a significant factor – high-profile celebrities talking about their new-found sense of purpose and fulfilment. Films, articles and books emphasising the wonder and joy babies bring into our lives. In the premier May 2000 issue of *Aura* magazine for the more mature woman, even celebrated feminist Germaine Greer admitted: 'The truth is . . . I was desperate for a baby and I have the medical bills to prove it.'

It's human nature to be influenced by the pressures and trends you see around you. Problems are likely, though, when the decision to mother is made in immediate response to them and not after thought and deliberation. Another way of thinking about this is, if you are so easily swayed by the opinions and actions of others, are you ready for the responsibilities of motherhood?

Wanting to experience pregnancy was also a motivating factor for some. A woman's body is designed to have babies. It is something we can do. We have hormones which can stimulate our desire to have babies. In a compelling BBC 1 documentary on surrogate motherhood, screened in early 2000, a three-times surrogate mother confessed how much she loved being pregnant, even though the prospect of raising a child did not appeal to her. She didn't want children of her own, but had to fulfil a need in her to experience pregnancy and childbirth.

Creating another life can feel like a miracle. It can give a woman with low self-esteem a sense of great power. I spoke to many mothers who loved being pregnant.

The vital question here is, how prepared is the woman after childbirth for the constant pressures of child-rearing?

Another reason for having children was a fear of loneliness. Women who have lost their parents, or those who don't feel close to their family, admitted that this may have been one of the motivating factors. Some were honest enough to say that they feared a lonely old age.

Some women said, as men are often thought to say, that they wanted to have children as a way of leaving something of themselves behind – making sure that their genetic line would continue. Children are the link to the future. They offer a kind of immortality, a record of one's existence. The family dynasty carries on.

Many of the reasons given here may appear dubious to you. It seems obvious that they potentially create unhealthy situations for both mother and child. You should not have a child just in case you might regret not having one in the future. A child cannot guarantee that a woman is suddenly going to get self-esteem or become mature and responsible. Becoming a confident adult takes time. It is not achieved by the act of giving birth. A child should not be used to ease a parent's loneliness or to bring couples together. Nor should children be expected to take care of you in your old age, or to be loaded down with guilt if they don't want to keep the family name.

But before you judge too quickly, remember that, as questionable as some of the reasons are, they are also very human. No situation in life is straightforward, especially one as crucial as the 'mother-or-not' decision. Conflicting, selfish, difficult feelings are a fact of life. They often coexist with more positive feelings.

If you are seriously considering motherhood, and can relate to some of the questionable motives here, don't think that there is something wrong with you and you shouldn't have a baby. Nothing in life is ever black and white. It just means that these issues need to be worked through and considered before the baby decision is made. Chapter 5 will explore further the question of emotional readiness to mother.

## Making the decision not to mother

> 'I remember clearly the day I decided I didn't want children. It was Christmas Day, 1996, and I haven't changed my mind since. I spent it with my sister and her children. She had invited round her best friend who had children too. It struck me how miserable everyone was. The children kept moaning all the time. The parents didn't talk to each other, except about the children. I wanted to go out, but we didn't even go for a walk. We just sat watching television all day. It was claustrophobic, unstimulating and horrible.' (Tiffany, age thirty-six)

Some of the child-free women I spoke to said that feminist thinking hadn't played a part in their decision, but it's impossible to talk about the decision to have a child without mentioning the influence of feminism. Thanks to the feminist movement we have contraception, easier abortion, and access to careers and education.

We can see satisfying alternatives to motherhood.

Groundbreaking feminist texts, like those by Simone de Beauvoir, and books like *Baby Trap* by Ellen Peck, did loom large in many of the discussions I had with child-free women. Feminist thinking has made an enormous contribution to the motherhood debate. It has exposed the highs and lows of every aspect of a woman's life. It has highlighted the drudgery that comes with motherhood and validated the option of the child-free lifestyle. It has examined motherhood in depth. It has taught us to ask questions, to explore alternatives, to be ourselves and not just to go with the flow. It has taught us that we do have choices. It has given us permission not to feel guilty for the decisions we make.

Many of the women I spoke to who had firmly decided against children expressed how reaching a decision was a real relief. There was no more guilt.

> Jane had her tubes tied when she was thirty-six. 'After all that uncertainty, I felt like a burden had been lifted. All of a sudden I was free. Free to live my life. It was and is so liberating.'

> Caroline was only twenty-five when she made her decision. Like many of the child-free women I spoke to, she said that she had always known she was not the maternal type.
>
> 'I never could imagine myself being a mother. I didn't want to have a baby. I don't think I ever wanted to have a baby.' A dedicated musician who performs all around the world, family life never appealed to her. 'When my mother told me she hoped I would marry young like her and have three kids before I was thirty, I felt dreadful. This was not the life I wanted. I couldn't imagine how anyone would want it.'

Not feeling maternal was often mentioned. 'I never had any interest in babies at all. I still haven't,' says Sally, age thirty-three.

Other women were profoundly disturbed by the actual spectacle of pregnancy and childbirth. For a generation of women who have only just begun to enjoy having control over their own bodies and lives, pregnancy repulsion could be seen as a reaction to the fear of loss of control.

> 'Pregnancy and childbirth may be one of the major events in a woman's life,' Linda, age thirty-nine, told me, 'but the whole thing repulses me. I don't ever want to experience such pain. To feel so vulnerable and helpless and at the mercy of the unpredictable hand of nature.'

Other women hated the idea of something alien growing inside them. Words used included 'growth', 'parasite' and 'thing'. The baby is seen not as a miracle of new life, but as a hostile intruder. Some said that they couldn't bear the thought of their bodies being tortured and torn. Weight-gain was also a big concern, and looking bloated. Others thought that having a baby would make them less sexy. That they would lose their partners when they got pregnant, that maternity and sex didn't go together.

Not feeling maternal and anxiety about pregnancy and childbirth were common themes, but the significance of influential people in the child-free decision also figured strongly in discussions. Most often mentioned were partners, other child-free women, mothers and family members.

Women in committed relationships said that they made the decision together with their partner. His or her views were often crucial. If both partners are of like mind and don't want children, bonds are often strengthened. Sometimes, though, when one partner is reluctant and the other not totally committed to the child-free lifestyle, there can be great pain and conflict.

In some cases the experience of a partner's instability was a powerful event that initiated the child-free decision. A painful breakup or divorce undermined faith in domestic bliss. For Ann, age forty-two, the realisation that a man would not always be there for her led her to conclude that self-reliance was the only reasonable alternative.

> 'I got seriously involved and thought, this is it. Marriage

> and kids. Three years later the relationship ended badly. Looking back I knew it was good to break up. We weren't right for each other. But I can also think that the pain it caused me at the time made me realise that my happy family dreams were unrealistic.'

The decision to forgo children was sometimes based on a rejection of the notion of love and commitment or disillusionment with it, either through fear of its impermanence, as in Ann's case, or a fear of commitment to another person. Unwilling to take on the responsibilities of parenthood outside a relationship, these women have learned from experience that long-term commitment to rear a child is an unrealistic expectation.

Not being in a relationship that is conducive to having children was frequently mentioned as a reason for remaining child-free, and this will be explored in more detail in Chapter 8.

Several child-free women mentioned having met a woman without children who lived her life in an inspiring, positive manner. There seemed to be no shortage of positive role models for women without children, especially among lesbian couples.

Women with children sometimes had a negative influence. The isolation and frustration of motherhood tended to steer some women away from having babies of their own.

> 'I look at housewives, and all I see is boredom, isolation, exclusion and a lack of direction and purpose. Every time I'm with a group of mums I'm so glad I'm not one of them. I don't get any intellectual stimulation. I see the effect having kids has on their figures, their careers, their marriages, their self-worth and their finances. It's horrible.' (Susan, age thirty-four)

Disappointing and difficult relationships in a woman's family of origin, and the emotional wounds they cause, also figured strongly in discussions about the child-free choice. The most influential relationship of all, it seems, is with one's own mother. For many child-free women their mothers' lives were an example of how they didn't want to lead their lives. In America between 1982 and 1988, cultural historian Shere Hite surveyed almost a thousand women and found that over 80 per cent had a staggering fear of becoming like their mothers.

Our primary source of knowledge about motherhood comes from our mothers. We look at our mothers and ask questions. Did she enjoy being a mother? How fulfilled was she? Did she have a career? Did she give it up for us? Was she often lonely, bored and frustrated? Did she stay in an unhappy marriage for us?

All these questions are crucial for women who make the child-free decision. They have often concluded from their own mothers' lives that having children was not the passport to fulfilment and wholeness, that there is so much more to life than marriage and children.

Many mothers, even today, give up their careers to have children or stay in unhappy marriages 'for the sake of the children'. Child-free women may have heard their mothers telling them how much they enjoyed being mothers and didn't mind the sacrifice, but they could see for themselves the frustration and resentment, the loss of identity.

In what has now become a classic study, *The Captive Wife: The Conflicts of Housebound Mothers* (Routledge, New York, 1984) sociologist Hannah Gavron interviewed a large number of mothers in the 1960s. She concluded that women experience isolation, insecurity and a loss of confidence when they have children.

Since the 1970s the mother–daughter relationship has received a great deal of attention. There is a real need for bridges to be built between the generations, between women who have different life choices. A healthy understanding between mother and daughter can pave the way for more honesty and frankness between women. Less attention has been paid to the influence of the father figure in a woman's life. But clearly a difficult or distant relationship with the father can also leave women with an emotional emptiness, a void that leads to an inability to want to nurture or have children of their own.

Caring for elderly parents or other family members could also make a woman decide not to have children. Sandy, age thirty-nine, took care of her father for ten years when he got ill and frail. 'I simply didn't have the time or energy to have my own family. And now that my father has gone I just want my freedom.'

In some cases the decision to remain child-free was career-based.

> 'Theatre production and motherhood just don't go together,' says Karen, age thirty-seven. 'Last year I was in the States for six months teaching and the rest of the time in Europe directing. I adore my work. I love being part of a team. I love the chaotic, late hours. Children wouldn't fit into my life.'

Until the career ladder became available to women, having children and raising them was the only real way women could find meaning and purpose in their lives. There were exceptions, of course. A few remarkable women defied the norm, but on the whole we didn't have a choice. Now we do. Family life does not have to take centre stage any more.

Study after study shows that combining motherhood with career is a challenging juggling act. A new survey conducted for BBC *Panorama* revealed that a third of working mums quit their full-time jobs for part-time work or give up their careers altogether.

'How mummy misses top jobs,' writes journalist Deborah Collcutt. 'Women are losing out on promotion because bosses still believe they are on a mummy track and do not offer such good value for money as men' (*Daily Mail*, 17 March 2000, p. 41).

Women struggle to balance career and family, and in most cases career suffers. The lack of flexibility in the workplace, the high cost of childcare, inadequate childcare facilities, and feelings of guilt about leaving the child place an incredible strain on mothers, many of whom find it impossible to cope. Some women bypass these stresses by choosing not to have children.

> 'I have seen how distracted working mums can be when their child gets ill or they can't sort out daycare,' says Hilary, age thirty-two. 'They can't concentrate on their work and they

> expect everyone else to be sympathetic. I don't want to jeopardise my career like that.'

Many women in their thirties reach a point in their careers when decisions need to be made. It is no accident that the mid-life career crisis is often coincidental with the mortality of fertility. Because the workplace has still not found a way to incorporate and embrace mothers, the baby decision has a real black and white feeling about it, creating a subtle but intense pressure.

Generally, though, it is unfair to say that a child-free woman makes a straightforward choice between career and children. This is popularly thought to be the explanation for the woman's child-free status, as seen in the stereotype of the bitter, neurotic career woman who feels empty and unfulfilled at menopause. But from my discussions, it became apparent that although career considerations did reinforce a woman's decision not to have children, the means by which the decision is reached were usually far more complex.

> Janet is a 46-year-old managing director. 'If I'd wanted children, I would have had them and somehow managed. It's easier to let people think that my career came first, but the truth is my mother's slow and painful death from cancer gave me a subconscious desire not to repeat the pattern and inflict the same pain and emotional trauma on a child when a parent dies.'

Many of the women I spoke to who were child-free said that work wasn't the sole centre of their existence. Not all of them were high-flyers with demanding careers. Many were in lower status or part-time jobs or were even housewives. Money was occasionally, but rarely, cited as a reason for not having children. If money was an issue it centred more on the fear of poverty, of not being able to feed and clothe a child, rather than a desire for wealth. Others said that not having children had liberated them from the necessity of earning high salaries and having material goals.

Far more frequently mentioned than work or money was the woman's desire for freedom, freedom to pursue her interests and her life without feeling that it needed to be validated by children.

> 'I don't think I would have been suited to being a mother,' says actress Stephanie Powers. 'It's a big job, and there are too many other things I wanted to do in my life. You have to sacrifice a tremendous amount when you have kids, and I wasn't willing to do that. I'm just being realistic when I say I didn't have the guts.' (Stephanie Powers, age forty-seven, quoted from *Mail on Sunday*, 23 April 2000, p. 25)

Child-free women, free to live their lives in the way that they choose, often find themselves accused of being selfish. In professions where physical beauty is highly rated, some women frankly admit, as *Dallas* star Victoria Principal did to *Redbook* magazine, that pregnancy is not in their best physical interests.

On 7 March 2000 the Bishop of Rochester, responding to growing evidence that more women are not having babies, sparked a fierce controversy when he denounced couples who didn't want children as selfish and dangerous. He asserted it was wrong to opt out of family life for the desire for more comfort, freedom or travel. The Right Rev. Michael Nazir-Ali went on to describe children as 'a basic good of marriage and not an optional extra' and spoke of 'an age of excessive self-regard'.

In response many child-free women suggested it was about time women started putting their own needs first and stopped losing their identity in the family. They argued that this wasn't selfishness, but a healthy search for self-identity. Some hinted that it is women who mother who are the more selfish ones.

> 'I see so many children born into unhappy marriages and broken homes. I think that's cruel and thoughtless.' (Karen, age thirty-six)

> 'A lot of women have kids for the wrong reasons. Isn't it selfish

to have a baby for the wrong reasons or if you are not sure?' (Amanda, age forty)

'The world is overpopulated enough. I think mothers are thoughtless because they expect other people to put up with the bad behaviour of their children.' (Lucy, age thirty-nine)

'There are enough unwanted and abused children out there within the "sanctity" of marriage. Perhaps it is a selfish decision that we made. We like our lifestyle and are happy with just the two of us. But what's wrong with that? Not having children doesn't make our marriage any less worthwhile.' (Linda, age thirty)

Many childless women felt that the Bishop of Rochester could hardly be more wrong. Rather than being self-indulgent they felt their attitude was responsible. They saw couples with children often expecting the entire world to revolve around their offspring. They resented the deluge of rules about maternity and paternity rights, and the constant demand of mothers for flexible working around children which they themselves had to put up with. They felt that people with new babies were boringly self-absorbed.

Some said that the determination to remain childless, rather than being selfish was a sign of devotion to a partner: the union is so happy that it needs nothing more. They condemned the attempt to have children to hold together a rocky relationship. Many argued that the crisis in modern British society was not over childlessness but the irresponsibility and self-gratification attached to child-rearing. Children appear to be regarded as little more than fashionable accessories with no recognition of the enormous duties and demands they entail.

Selfish or not, the woman who decides not to have children doesn't feel comfortable with the notion of devoting her life to a family. Her strongest desire is to have an identity outside the traditional family role. She does not want to lose her identity in motherhood. This is so vital to her that she is prepared to say no to having children. For many women this is too high a price to pay, but for the child-free woman it is not. She may or may not have regrets, but what is crucial is that she is free to make the choice.

> 'Why should we be pressured into having a child and made to feel guilty if we don't? Surely it is everyone's fundamental right to choose whether they see fit to have children.' (Linda, age thirty)

Those who decide against motherhood often conclude that children will entail greater costs than rewards. Some think they can't create a situation suitable for childbearing, or that the responsibilities of parenthood would have disastrous consequences on other valued aspects of their life. The potential costs of motherhood are not worth the risk – and having a child is certainly a risk. When a woman has her first child she just doesn't know what the actual experience of motherhood will be. And for some women this is just too unpredictable a choice to make.

> 'Maybe I'm just afraid of having kids,' says Linda, age thirty-nine. 'I don't know anything about babies. Maybe that's just an excuse. Maybe it comes down to learning to deal with being a mother, throwing myself into the great unknown, because if I really wanted a baby, I would find a solution.'

## Childless, but not by choice

> 'I thought everything would come to me,' reflects Mandy, age thirty-nine. 'As women we were told we should "go out and do what you want to do". I didn't know what I wanted to do in my twenties, so I didn't do much really. Then in my thirties I thought I must get a career. I got married and delayed children for a few years. That's what's so scary. I'm trying to have children now. I keep asking myself what the hell I did with my twenties and thirties. I thought I'd just get pregnant really

> quickly, but when I had a miscarriage I realised it's not going to be that easy. I have to face the fact that I could be infertile by now.'

Being childless by choice is very different from being childless but not by choice. When so much female identity rests on motherhood, the woman who does not have children by choice can at least reinforce her femaleness by saying, 'I could have had a child, if I wanted to.' The infertile woman has the choice taken away from her.

Many infertile women live with a pain and an anguish that are acute. Some have postponed the decision to mother for too long. They are approaching menopause and their chances of conceiving are fading fast. Infertility comes as a devastating irony after using years of contraception. Others start trying in good time, but find that for various physical reasons they can't get pregnant. Either way, infertility comes as a terrible blow.

When the decision to mother has been made after several years of postponement and a woman has problems conceiving, there is often shock and disbelief. A sense of disempowerment may set in, especially if she felt that her life was under her control. How could this happen? Why has this happened? This isn't how I planned my life to be! What am I going to do with the rest of my life? It is easy for her to start blaming herself. If only she had thought about having a baby sooner. If only she hadn't been so wrapped up in other things. If only she had taken better care of her health. If only . . .

She may go through a period of denial, refusing to believe that this could be happening to her. Expensive fertility treatments may become the sole focus of her life. Eventually, though, after shock, disbelief, self-hatred and denial, she will have to come face to face with her grief and disappointment. If infertility is the case she must learn to deal with the decision nature has made for her.

Choice is a privilege that not all of us can enjoy. One in six couples experience problems with fertility. No matter how liberated, advanced and in control of our lives we may feel that we are, some things are out of our control. The biggest challenge now is to come to terms with the sense of loss and reinvent our lives in a fulfilling way.

Infertility is such a huge topic that Chapters 9 and 10 have been devoted to it.

## Decision by no decision

'The new Scarlett O'Hara is the high achieving woman who tells herself, "Oh fiddledeedee, I'll just think about getting pregnant tomorrow" ' (Gail Sheehy, *The Silent Passage* (Random House, New York, 1991), p. 104).

Some of us, unable to decide, deal with the anxiety by postponing the decision until circumstances make up our minds for us, or menopause arrives and makes the final decision.

American psychologist Mardy Ireland in her book *Reconceiving Women* (Guildford, New York, 1993) describes women who postpone the baby decision as 'transitional women'. She describes them as women who want to pursue social and career opportunities but also want to think about a family. Many of them have specific criteria that need to be fulfilled before they start a family. Others have no such agenda and are simply living their lives one day at a time, believing that someday they will have a child. In the meantime other goals, personal, educational, financial and so on, are more important.

Women are becoming more and more cautious, nervous and critical. We want to wait until the time is right. Until we have financial solvency. Until careers are more certain. Until decisions about work and day-care have been made. Until we can come to terms with the thought of our bodies changing. Until we have a clearer idea about what kind of mother we will be. Until we are sure that we will be good mothers. Until we are sure that we have our own identity before we join the mummy club. Until we stop wanting to go to wild parties and spur-of-the-moment vacations. Until we are in a committed relationship. Until we are really convinced that we are ready.

Waiting for that perfect time, it is easy to put off the decision about motherhood year after year – until suddenly there seems to be little time left. This is what happened to Lucy, age forty-three.

> 'I was really enjoying my job so I didn't want to stop working to have a baby. There were just so many exciting opportunities for me. My hours were unpredictable, and I knew that having a child would get in the way. Then we decided to sell our flat and buy a house. It took ages to find the right place, and just when we finally had, I was offered a job in London and we decided to move there instead. Property is expensive in London, so at first we rented a flat. By now having a baby just seemed an impossible idea. I saw my friends struggling with the demands of childcare and a career, and I didn't feel ready to face all that. My mother was also very ill at the time, and it was hard just making sure that she got the care she needed. I didn't stop using contraception until I was forty. Nothing happened. I probably just left it too late. A part of me is relieved. I'm not driven enough to go for fertility treatment. It's unlikely at my age that I will get pregnant, but if I did I would do my best to be a good mother.'

The average age for first births is edging upwards towards thirty. Birth rates have increased dramatically in the 35–45 age group. Postponement of motherhood has become a way of life for a whole new generation of women. But somewhere along the line, postponement turns into a decision.

If you are in your forties and still undecided, you may well join the increasing number of women who are choosing to say no to motherhood. If you want to achieve certain goals before you get pregnant and want to wait until you are ready, you may find that you never feel ready. With your biological clock ticking you realise a choice has to be made.

If you are unsure about children you may make a passive choice. You simply stop using contraception and adopt a 'wait and see' attitude. Should you get pregnant without trying, this can cause much soul-searching. You may find that you both want and do not want the baby.

Most of us, however, prefer to make a conscious choice about pregnancy, but there are often fears which prevent us making that choice. Fear of finding the right partner. Fear of how a baby will impact career. Fear of no longer being physically attractive. Fear of never losing the weight. Fear of losing identity and so on. All of these fears need to be addressed and addressing them takes time. But the problem with taking your time to feel emotionally and physically ready is – yes, you guessed it – the biological clock.

If you feel ambivalent about children, gambling with fertility as you make your mind up does not often ease the anxiety. It just delays it until the ticking of the biological clock gets so loud that you have to make up your mind about what you really want. A day comes when you have to ask yourself, 'Do I want a baby or not?'

Hopefully, by weighing up the pros and cons, you can answer this question before the clock stops ticking and you run out of options. Hopefully you will be able to understand the reasons you have for delaying the baby decision. And hopefully, with the help of this book, you will be able to make a life choice which makes sense to you as an individual.

Now that we have looked at the decision-making process, it's time to briefly turn our attention to the consequences of those decisions. The intention of the next chapter is to help you feel better informed if you are still undecided about motherhood. We'll take a look at the 'for better' and the 'for worse' of life on both sides of the fence – how having a baby can change your life, and how delaying, or not having, a baby can shape your life.

# 4

# Life Choices

> *Whether we decide to try to conceive or not, to continue a pregnancy or not, to become a mother or not, these choices have consequences that profoundly affect our bodies, our minds, and our sense of personal possibilities.*
> (Dianne Hales, *Just Like a Woman* (Virago, London, 1999), p. 178)

Today you can choose to become a mother or not. You can find out if you are pregnant at a very early stage in gestation. You can terminate or continue the pregnancy. What you can't find out, though, is how the life choices you make will affect you. This chapter is an attempt to give you some insight.

## Pregnancy and childbirth

> 'There is something both wonderful and terrifying about being pregnant. It's a bit like moving house. For a while you are in no-man's-land. Neither one place or another. There is excitement and anticipation, but also a sense of loss because you are leaving something behind. For months you are busy with

preparations. And when you arrive you are never fully prepared.' (Michelle, age twenty-nine)

The only thing you can really expect when you are expecting is to feel transformed both inside and outside. Rapidly changing moods and heightened sensitivity reflect the millions of changes within your body and the constant adjustments it has to make. You will gain weight. Your heart beat will speed up. Pregnancy hormones will flood your system. Your pulse will race faster and your body temperature will increase. Your joints will loosen and your centre of gravity will change. Your stomach will be lifted and the position of your heart will change. Having a baby is one of the most enormous physical tasks your body will ever accomplish.

Pregnancy links mother and child in a most intimate way. The fetus shares your blood, and eats your food. But the fetus also puts its own needs first. The goal of pregnancy is the development of a healthy baby. The physical and emotional well-being of the mother takes second place. For every woman who enjoys taking a back seat to experience the miracle of new life growing inside her, there is another who finds the loss of control over her body profoundly disturbing. The nausea, the weight gain, the shortness of breath, the fatigue make some women feel as if they are selflessly indulging a needy parasite.

Levels of physical and emotional discomfort during pregnancy vary from woman to woman, but there isn't a pregnant woman alive who wouldn't admit to feeling anxious at times. She is unlikely to complain though. After all, shouldn't she be happy?

However delighted or anxious she is, a pregnant woman will find herself caught between the idealised stereotype of what she is supposed to think and feel and how she actually feels. For some women pregnancy does represent fulfilment, femininity and power. They literally glow with beauty and pride. 'Yet,' writes Susan Maushart, 'for every woman who revels, goddess-like in her pregnant body, there is another who feels downright grotesque' (Susan Maushart, *The Mask of Motherhood* (Random House, Australia Pty, Ltd, 1997), p. 55).

Childbirth, like pregnancy, is another completely mysterious, thrilling, terrible, bewildering experience few of us are prepared for. Childbirth classes 'teach' us how to manage the pain – which is a big step forward from the days when women had absolutely 'no idea' what to expect – but for the great majority giving birth 'is nothing like the books say'.

There is nothing predictable about childbirth at all. Many women emerge from the delivery ward feeling shaken, betrayed and amazed at how ignorant they were. Nothing can prepare you for the shock. Nothing can prepare you for the pain. Nothing can prepare you for the exhilaration. Nothing so thoroughly shatters your conviction that you have some kind of control.

Some bodies will be more equal to the task of pregnancy and childbirth than others. But this is something you can only find out should you actually get pregnant and go into labour. From conception onward having a child is a journal of new experiences. Every day brings new surprises. There are no certainties any more. It's frightening not really knowing what to expect.

Even more scary, however, is knowing that after childbirth there are no more open options. The decision to mother has been made. There is no going back.

## Becoming a mother

> 'I had this idea in my head of what being a mother would be like. I'm an organised person and I thought I could manage my career and motherhood, no problem. But babies have their own schedule and it's not the same as yours. They are so unpredictable. I didn't realise how much they disrupt your life.' (Nicola, age thirty-six)

The reality of life after birth can be quite different from what many women expect. Few women really know what they are letting themselves in for. There is so much fake sentimentality surrounding motherhood and the truth is that our generation of women is simply not well prepared for motherhood.

The first months after childbirth are a time of great vulnerability. Not only is the body working incredibly hard to return to the pre-pregnant state, but with the massive hormonal shifts taking place emotions are in turmoil too. Doctors tell mothers to expect to feel disorientated for a few weeks. What they don't say is that for many women it takes months or even years to get their bearings. Most eventually regain a sense of balance, but a few find themselves in the depths of despair.

For many new mothers the biggest shock is not just the physical and emotional readjustment to a baby, but finding out that motherhood doesn't come naturally. Things don't just fall into place. Maternal instinct doesn't naturally arrive the minute you give birth. I spoke to over fifty mums when writing this book, and almost all agreed that they had underestimated how difficult being a mother was. Such a thing as a 'natural' mother does exist, but she is in the minority. Many women feel that they don't have much experience with babies, can't trust their instincts and don't really know what to do.

Breastfeeding, for instance, can be a wonderful bonding experience, but it can also be a source of stress. In some cases it just simply doesn't suit either mother or child. Few women anticipate such a so-called natural event ever being problematic, but it certainly can be. Likewise bonding with baby may also not instantly occur. That gush of love, joy and exhilaration a woman is supposed to feel when she gives birth may be strangely elusive. Building a relationship with the baby takes time.

Many mothers constantly ask themselves if they are doing a good job. To a large extent women imprison themselves in a host of assumptions about 'good' mothering and the maternal instinct. Mothers are the most self-critical of women. Few really accept that it is OK to fall short at times – that it doesn't automatically mean becoming a failure as a mother. It is hard to grasp that the sentimental notion of unconditional maternal love is an impossible ideal. That just because you feel opposing pulls between motherhood and your own needs does not mean you cannot be a good parent.

Mothers are supposed to be serene, but looking after a baby actually produces some of the most intense feelings of panic, fear, rage and despair a woman can ever experience. For many women maternal skills have to be learned; they are not inborn. The feeling of being totally bewildered and weighed down by motherhood is a commonplace experience.

In her book, *Life After Birth: What Even Your Friends Won't Tell You about Motherhood* (Viking, London, 2000), author and mother Kate Figes discusses how inadequate she felt in the early years of motherhood. She describes how she couldn't think straight or feel useful. How exhausting the constant demands of the newborn were, and how chronically sleep-deprived she felt.

Mothers everywhere continue to hope that one day their lives will get back under control but they soon learn that, however hard they try, they can't go on as normal. Everything in their life changes. Some said that they felt buried alive by all the changes and their new responsibilities. Others said they were coping. Others said they felt reborn. Whatever the case, one thing was clear: all aspects of their lives – their self-esteem, their bodies, their careers, their relationships, their social life – are touched. In the words of German ethnologist Leo Frobenius, 'A man spends a night by a woman and goes away. His life and his body are always the same. The woman conceives. As a mother, she is another person than the woman without the child' (Dianne Hales, *Just Like a Woman* (Virago, 1999), p. 196).

Studies on the transition to motherhood reveal that, although some women do grow in self-esteem and confidence, far more experience a loss of identity when they become mothers.

> 'What the hell had I done with my life?' says Sandy, age thirty-two. 'I used to have a good job. Now with a tiny baby I was shattered, unable to concentrate on anything for a few minutes. I felt so alone. I felt as if I'd dropped out of the human race. I was isolated, disorientated and tired. The chaos never lifted.'

If a mother is at home full time, caring for a baby can be lonely and unstimulating. If she works she will find that her priorities have to change. Whether working or not working, life doesn't revolve around her needs any more. It revolves around the baby, and a woman can feel as if she is losing her whole life.

A mother's body changes when she has a baby. Most mothers do tend to put on extra weight (around 5 to 8lb on average), and breasts, stomach and feet will probably alter in shape. This does not necessarily mean that a woman's body will change for the worse, but it does mean it will be different. How much it changes though, and how it changes, depends on the individual and her lifestyle. Many mums said that having a baby made them feel less inhibited about their bodies. Some said it made them more appreciative of their bodies and that they had never felt so attractive. On the other hand some also said that they felt older, heavier and less attractive after childbirth.

However hard a mother tries she often finds it a struggle to balance the demands of career and family. Many mums worry about the consequences of having a baby for their career development. Others say that their employers have little sympathy or understanding.

Many mothers told me how frustrating it was that they couldn't give their careers the same kind of focus and commitment. A parent's timetable is very much dependent on a child. The loss of flexibility often compromises career development. A survey conducted by BBC *Panorama* highlighted the plight of many working mums. It revealed that a third of working mums are quitting their full-time jobs or giving up work altogether.

It is only by returning to work full time following the birth of a child that women are able to stay on track with their career and earnings, but lack of flexibility in the workplace can put unbearable strain on women trying to juggle the demands of work and family.

Working patterns tend to alter when baby arrives and for many there is often an entire career rethink. Some start new careers. Some decide to work from home or to work part-time. Some mothers stop working altogether. Others find that, even though it is tiring and difficult to juggle both home and work, work is important for their self-esteem. Some want to work, but can't find or afford suitable childcare. Others resent the fact that they have to leave their babies and go back to work for financial necessity.

Whether working or staying at home, it was clear that feelings of guilt about working or not working figured strongly for both stay-at-home and working mums. Stay-at-home mums often felt guilty for not bringing in an income and frustrated by the lack of stimulation and challenge. Working mums often felt overwhelmed by the demands on their time. Few expected to feel so conflicted: on the one hand, relieved to be able to get their lives back on track, on the other hand, regretful, guilty and torn about not being there for their child.

A recent spate of novels have dramatised what is becoming a common conflict between the demands of work and the demands of home. In Diana Appleyard's *Homing Instinct* (Black Swan, 1999, p. 230), attempting to surprise her children, career woman and thirty-something mother Carrie arrives too late at the nursery, and sees them being collected by their nanny. She sees a snapshot of her children's daily life where she plays no part. She feels 'neither needed or wanted. A mother surplus to her children's requirements'.

Another big change babies bring is financial. Having a baby is expensive. Depending on the income levels of the parents, a mother may find herself forced to work longer hours, to economise and budget as never before. And the bad news is that the financial burden doesn't ease as the years go by. Children get more expensive as they get older. Many mothers work hard and sacrifice much to give their children the advantages they think they should have.

Relationships with partners change. Now there are three to consider. Some mums said that the baby made the bond grow stronger. Others would agree more with Nora Ephron: 'A baby is a hand grenade tossed into a marriage.' Many women said that they did experience difficulties adjusting after the birth of their first child. Lack of togetherness, resentment, disagreements about parenting styles, fatigue, increased responsibility, and the lack of time, privacy and energy for satisfying sex can strain any relationship. Loss of libido after childbirth was also common.

When a child arrives on the scene a mother may not be prepared for how much the loss of personal space will affect her. Privacy and having enough time for herself are things of the past. A woman accustomed to taking her freedom and autonomy for granted may find it claustrophobic to have to work her life around the needs of another. Many mothers, much as they love their children, do feel trapped by them. On the other hand others said that they loved the responsibility and closeness of caring for a child twenty-four hours a day. It made them more compassionate and less self-centred.

Of all the many changes motherhood brings, it seems that by far the biggest adjustment is the new and disturbing loss of control. Few women are prepared for how much love and responsibility they do feel for their children. But along with this love comes a terrible fear for their safety and well-being, which can destroy peace of mind. There are of course exceptions, but even the most tough, controlled and sophisticated of mothers will discover that, in the words of Dianne Hales, 'To become a mother is to put your soul on the block . . . to open yourself up to bartering and blackmail, to discover yourself – sometimes much to your own amazement – acting just as mothers always have' (*Just Like a Woman* (Virago, London, 1999), p. 212).

Putting the needs of your child above your own. Worrying constantly about your child's well-being. Thinking with your heart and not with your head. Some women enjoy their children so much every sacrifice is worth it. Others love their children but struggle more to balance their needs alongside those of their child. Ambivalent feelings about undertaking such a selfless task are not uncommon, even though few would dare admit to them.

Every woman's experience of motherhood will be different. It is impossible to know how you will be affected. It depends on so many constantly changing factors. Your life history, your social background, your family, your lifestyle, and how you choose to mother will all influence your experience of motherhood. The job of mothering is a constant work in progress. The only thing that can be said with any certainty at all is that a baby will reorganise your life completely. There will be continual change around the developing child. In much the same way that your body has to change and stretch to nurture a child, so your life has to accommodate itself around baby. Nothing in your life is quite the same again. And even the most committed, focused and calm among us may find the irreversible unpredictability of all these changes quite daunting.

If a woman can adjust to the constant change in her life and philosophically accept both the good and the bad, she often finds the positive does outweigh the negative.

> 'I can't imagine any other experience in life being so incredible and fulfilling. Those wonderful special moments of closeness when you cuddle up with your baby. The joy of watching your child grow and develop. The fun you have as you view the world once again through a child's eyes. The feeling that what you are doing is completely and utterly right.' (Sally, age thirty-six)

For Sally, like the majority of the mothers I spoke to, there is nothing more gratifying and incredible than having children. Words cannot really describe how much joy and wonder a child can bring into your life. Mothering isn't easy, but it yields rich rewards. Through having her own child a woman discovers a depth of love, compassion and fulfilment that enriches her life.

Much depends on how well you can cope with your vulnerabilities and how willing you are to adjust to the enormous changes a baby brings. You may feel confident that motherhood can and will enrich your life beyond measure, or you may not feel so sure. You may decide that you don't want to stretch your body and your life to accommodate a child.

## Choosing to say no

> 'I don't want to have a baby. I got pregnant when I changed the pill. All I could think was "I don't want this to happen to me." I was scared and horrified. I would have done anything, even

> reached for the knitting needles, to get rid of it.' (Michelle, age forty)

After the baby boom, we have the baby dearth. Fewer babies were born in the UK in 1999 than in any year since the Second World War. Part of this is due to the increase in fertility problems – the trauma of infertility will be discussed in Chapter 9 – but it is also due to the fact that more and more women are choosing to say no to children.

Every time you use a contraceptive you are choosing not to become a mother. A century ago the only way women could control their fertility was to abstain from sex. Almost all the child-free women I interviewed were thankful that they were given the right to choose, that they could confirm their decision on a daily, weekly or yearly basis.

Some of the horrific extremes women have gone to to control fertility illustrate just how strong the determination can be not to have a child. Until abortion became legal in 1967 acid drinks, razor blades, coat hangers or knitting needles were the last resort for desperate women. Thankfully today, should contraception fail, most of us can choose to have a safe abortion.

About a third of the child-free women I spoke to had resorted to abortion at some stage. For many of them the accidental pregnancy brought their feelings about motherhood into sharp focus. Decisions had to be made very quickly. The issue of motherhood or not must be confronted head on.

Even for those firmly committed to the child-free lifestyle, an abortion always brought dramatic mixed feelings to the surface. Some mentioned that they had suffered psychological distress after an abortion and worried they might regret it later on. The great majority, though, seemed thankful and relieved that their lives had not been ruined. They felt happy with the decision they made and felt it was right for them.

Almost all the women I interviewed who had decided not to have children said they wouldn't hesitate to have an abortion if they got pregnant. There was no way they would even contemplate having a baby. Others, who felt abortion was ethically wrong, said that they would have to put their child up for adoption. A late period could throw some women into considerable turmoil. They preferred sterilisation rather than the constant stress of worrying about getting pregnant.

Despite the fact that male sterilisation is a far simpler, quicker operation, more women, either single or in couples, have the operation than men. Sterilisation is fast becoming one of the most popular forms of contraception, even though it is very hard to obtain on the NHS. Fees for female sterilisation are just under £500. There are two procedures for female sterilisation. The first is called a salpingectomy and requires a general anaesthetic for part of the fallopian tubes to be cut. The second is a laparoscopic technique, which can be done under local anaesthetic to cauterise the tubes.

Many child-free women expressed their frustration about how difficult it is to find a doctor willing to perform a sterilisation. Women in their twenties are often told to wait until their thirties. And even then it may take a woman several years to convince a doctor she is serious in her conviction. Sterilisation is usually a permanent procedure, so in some ways it is sensible that a doctor encourages a woman to give it a lot of thought. Most of the women I spoke to who were sterilised didn't regret it, but some did.

Even though reversal is possible in about 40 per cent of cases, sterilisation is intended to be a final step. Being such a drastic decision for a woman to make, it will always generate interest and controversy. Some child-free women enjoy being unconventional and don't mind bringing up the subject. Others prefer not to tell anyone about it, especially not their family, for fear of upsetting them. They feel that it is kinder to let everyone think they are infertile.

For some women sterilisation is just too extreme a step to take. Others rejoice in the liberty it offers them. No longer do they have to worry and panic about possible pregnancies. They are free to lead a life without children.

## Life without children

> 'I know a lot of people think my life is incomplete without children. I wonder why they think that. I enjoy every aspect of my life. I think some people are just threatened by the idea that you can feel complete without children.' (Abigail, age thirty-five)

In many ways deciding not to have children is an even more scary and unpredictable life-choice than deciding to have them.

The woman without children is, in the words of Mardy Ireland, 'a trailblazer'. You place yourself on a little-travelled path. You have not got the comfort and familiarity around you of what a woman 'should' do and be. You have made a conscious decision to explore other avenues of expression for whatever maternal feelings you have. You develop your identity in other ways. You live in a world not quite yet able to adapt to or understand you.

Many of the women I spoke to felt happy and fulfilled by their lives and relationships. In fact studies show that stress levels among women without children tend to be lower than those with children. But, despite this, should you decide not to have children you may find that you have to defend yourself against our culture's lingering conception of the childless woman as unhappy, unwomanly, selfish, even dangerous.

However firm a child-free woman is in her convictions, she has not chosen an easy path to travel. Opting for such an unpopular choice means weathering social disapproval and personal doubt.

> 'I know I make a lot of people feel uncomfortable,' says Amanda, age thirty-eight. 'I'm not married and I don't want children. People either assume I'm desperate to have babies before it's too late or that I am obsessed with my career. It makes me angry sometimes. Just because I don't want a family doesn't mean something is wrong with me.'

Going against the norm can make you feel very vulnerable, depending on your level of self-esteem. If you aren't sure why you don't want children, you may at times feel that you aren't a 'proper' woman. You may feel that you are being selfish for wanting your freedom. It is very important that you understand why the choice you make is right for you, and come to terms with that choice. If you can't, then you are likely to feel anxious and conflicted.

Many of the women I spoke to said that coping strategies had to be developed. Consciously or unconsciously they try to muster network support by surrounding themselves with other single women or childless couples. Women without children do tend to think of themselves as unique. Some dislike it, but others quite like it. Very important to many of them is finding a place in the world where they could meet other childless women with whom they could identify. This could mean letting go of friends who become mothers and actively seeking out childless people.

It is easy to make assumptions that the child-free woman is often lonely, but this isn't necessarily true. Those who are single often have a rich and varied social life. Others find like-minded partners and value the intellectual and emotional closeness too greatly to take on the risk of a child.

Fears of loneliness can be powerful when making the child-free decision, but many of the women I spoke to understood that there are no guarantees in life. Having a child does not guarantee that he or she will be there in later life. Some suggested that mothers may in fact be more lonely in later life because of the void left by their children.

Many of the women emphasised their pride in their independence. They all rejected the argument that their identities as women would be jeopardised by not having children. 'I just don't need to do the mother thing to feel like a real woman,' says Sasha, age thirty-seven.

They did not believe they were denying themselves meaningful human relationships or that they were missing a vital stage in adult development. The rewards of personal accomplishments and autonomy superseded whatever pleasure children might bring. The costs of motherhood seemed to outweigh the benefits.

For those passionate about career, mothering did not promise to be rewarding enough to compensate for missed opportunities. Others said that they had watched the promising careers of other women take a nosedive when babies arrived. They felt they would not be able to combine work and child-rearing successfully and give the same level of commitment and focus.

It was rare for any interviewees to have no regrets at all about the child-free choice. A few vigorously asserted that they never for a second doubted that they had made the right choice, but the majority were happy to admit that having children was probably a fulfilling and wonderful thing to do. There was a certain amount of disillusionment with family life, but several mentioned that they often thought what life with children would be like, or even what their own child would look like. Almost all had not come to the child-free decision easily. There had been much thought and deliberation about the kind of life they could offer a child.

> 'Like every woman,' says actress and singer Elaine Paige, 'I've had a biological clock ticking away and it is a very strong, physical need. You can't deny that it's happening to you and it did make me consider very carefully the possibility of having children' (quoted from *Daily Mail*, 22 April 2000, p. 41).

Although some child-free women said they couldn't tolerate children at all, there were many who had just the opposite reaction and enjoyed cuddling and playing. But enjoying children was not sufficient reason for them to have one of their own.

Despite admitting to regrets and longings for what might have been, many were able to be rational about their decision. They felt that they would have even more regrets if they had a child. They focused not on what might have been, but on the reality of their lives now. The life they had chosen was the best alternative for them. They gathered strength from the conviction that they were making the right decision, whatever the pitfalls.

> 'Of course I have had my doubts. I look at families and I think I'm never going to experience that. But my life is rich in other ways. I have an active social life. I love my job. I really do enjoy my life. I don't want it to change in any way and it would have to if I had children.' (Abby, age thirty-five)

It was interesting to note that women without children were much more comfortable expressing regrets about their life choices than were mothers. I found it very rare to speak to a woman who revealed deep regrets about having children. Many felt that to admit regret would seem like a rejection of the child that they loved so much. They didn't mind admitting that they often felt ambivalent about motherhood and found many aspects of it frustrating, but talk of regret was strictly off limits.

The truth is that whatever life choice you make, there may always be a sense of loss and regret. If the choice has been made thoughtfully and carefully, you will be able to cope with difficult feelings when they arise. You may not be the perfect mother you thought you might be. You may have regrets about never having children, but, hopefully, making the decision to mother or not will have given you enough self-knowledge to know that the life choice you made suits you.

If you do choose not to have children it is important that you feel comfortable with your choice and the way you describe it. Many women prefer to be called 'child-free' rather than 'childless' because the latter implies a sense of loss, lack of choice and regret. 'Child-free', on the other hand, implies power and choice. Some women don't like 'child-free' because it makes them appear anti-children. Those who want to have children but can't often struggle with the term because they don't feel free at all. Generally, though, using the child-free description helps women who have chosen not to have children feel less victimised.

But what about the millions of women who can't make a choice? Those who postpone or delay making a decision because circumstance aren't right, or they don't know what they really want?

## Sitting on the fence

> 'I seem to be always making the decision. When I was twenty-five I said I'd have children when I was thirty. Now I'm thirty I want to wait another few years. I keep waiting for the moment when I know I'll be ready. It's quite depressing to think that the right moment may never come. I only hope I can live with the decision I am making. That I am doing the right thing.' (Sandra, age thirty-one)

In some ways keeping both options open is positive. You are always open to change. You have decided for now not to have babies, but accept that things might be different one day. The problem with continual postponement, though, is that it means you live in a state of division.

Sitting on the fence, trying to keep both options open, is stressful. The state of indecision takes its toll. When all options are left open you may experience emotional exhaustion and lack of direction. In addition it may be difficult to make important decisions in other areas of your life before resolving the motherhood dilemma. You may feel uncertain about your future, unwilling to take on long-term projects in your career and so on, because you aren't sure if you are going to have a child in the next year or two.

A number of women told me that they only realised how much of a drain their indecision had been after they had made up their minds, and they experienced a burst of energy and creative release.

If you have postponed the baby decision and then finally made up your mind to get pregnant because of pressure from the biological clock, you may be lucky and have no problems at all conceiving. On the other hand, fertility could well be an issue that you hadn't anticipated. Almost overnight your life may change dramatically.

If you are approaching forty, time can rush by at an alarming rate. It almost seems as if the biological clock picks up speed. Aware of declining fertility, you may start to count your life in months, or more precisely in ovulations. These cycles of time add a sense of stress and urgency to everything you do. You may find that your mood, your highs and lows, correspond to your cycle. Despair at the end of the cycle when you realise you are not pregnant. Hope and elation at the beginning when you think you may be. All of a sudden having a baby may become the most important thing in your life.

For every woman who finally makes a decision about whether or not to have a baby there is another who is still undecided. You may choose to deal with your indecision by ignoring it – the issue has become too charged to discuss with anyone or even to admit to yourself.

> 'It's something I really don't want to talk about any more,' says Kelly, age thirty-eight. 'The honest truth is I just don't know what I want. I keep hoping that one day I will wake up and know what to do. I just haven't got the time or the energy right now to think through all the issues and make up my mind.'

But ignoring the issue doesn't make it go away. It just delays it until the 'now or never age' when the biological deadline approaches, options start running out, and you have to confront the issue of whether or not you are ready to have a child.

# 5

# Am I Ready to Have a Child?

'When Taylor was born, it was the most mind-blowing thing. I thought, "Why did I wait so long?" ' (Dolores Oriordan, singer)

'I'm disgustingly happy. I'm obscenely enjoying my life (since the birth of my daughter).' (Caroline Quentin, actress)

'I'm in heaven, I was overcome with joy when I held my baby in my arms.' (Emma Thompson, actress)

The new millennium has begun with a haze of excitement about babies. When celebrities or prominent public figures have babies it eclipses everything else on the front page news. A new baby is seen to symbolise hope, new beginnings, all that is pure and good about humankind.

It is an irony that the 'have it all' generation of women have raised the significance of having babies to new heights. For all our liberation, the idea that having babies is what women do and what fulfils them is still strong. The cult of the baby is at an all time high. For successful career women in their thirties and forties a positive pregnancy test is hailed as the consummate achievement.

Yet in spite of this current baby craze there is no escaping the fact that fertility rates are gradually falling. More women than ever before are choosing not to have babies, or delaying the baby decision and taking risks with their fertility. For all the promised wonderland of motherhood, a growing number of us are not rushing into motherhood, but thinking very carefully about whether or not we really want to have a child.

But how do you know if you are ready to have a child? Remember as you think through the issues that there are no right answers. The purpose of this chapter is simply to help you focus your thoughts and think about what parenthood means to you.

## Be realistic

As mentioned in Chapter 4, the reality of being a mother can be quite different from the expectations and fantasies. Parenthood, unlike what some of us think, is not all effortless joy, fun and fulfilment.

> 'I thought taking maternity leave would be a bit like a vacation. That I would write my novel while my baby slept peacefully. That I would have time to exercise and get in shape. That I would feel a bond of love and closeness. But I was totally unprepared for the chaos. I didn't realise that a real baby doesn't just let you do all these things. They cry, they fuss, they demand. They make life incredibly hard.' (Ann, age thirty-eight)

Unrealistic fantasies and expectations about motherhood can be detrimental because they leave women unprepared for the reality. The more realistic your idea of motherhood is, the easier time you will have coping with the more difficult aspects of parenthood.

To help you recognise if your ideas of motherhood are based on fantasy or reality, ignore the media hype and gushing celebrity stories, and get as much information about being a real mother as you can. Read books and magazines about parenting and how it will change your life. Several are recommended in the Bibliography. Spend time with children so that you come face to face with the reality of being with them. Talk to other mothers about the realities of their lives. (But do remember that often mothers are overly anxious to present the positive side of parenting and may present an unrealistic picture.)

Remember that you owe it to yourself and your partner, if you have one, as well as to your future baby to make a decision based on reality not on fantasy, media hype, pressure from others or pressure from your biological clock. Having a baby is not something to be squeezed in before it is too late. Most important of all, try to base your decision on what having a child really entails. Be aware of both the potential gratifications as well as the potential frustrations of motherhood. As wonderful as they are, babies do not guarantee instant fulfilment. They are not paragons of perfection, but simply smaller, albeit cuddlier, versions of ourselves.

## Is the timing right?

You may at this point think you want a child, but you don't feel ready. If you are in doubt it is better to wait. Your biological clock may be ticking loudly, but you cannot afford to base your decision on this pressure. Your decision must not be based on biological-clock panic, but on whether or not you are ready to become a parent.

In the words of Beverly Engel, 'Becoming a parent for the first time is so stressful and requires such commitment that you must be financially, physically and most importantly emotionally ready for it' (*The Parenthood Decision* (Doubleday, New York, 1998), p. 86). Questions that Engel recommends potential mothers ask themselves to determine readiness include:

- Is my health good, both emotionally and physically, to withstand the stress and change a baby brings?
- If I am in a relationship, is the relationship secure enough to withstand the change?
- Do I have support from family or friends?
- Is my self-esteem high enough to cope with the physical changes of pregnancy?
- Do I understand what a baby is going to mean for my career?
- Can I realistically afford to support a child financially? Have I thought about life insurance?
- Do I have the time and energy for a child?

Determining if you are ready to become a mother involves assessing how much time and energy and money you have, whether or not your lifestyle is suitable and how emotionally ready you feel. If you will be a single mother there are special considerations, and if are in a partnership you need to decide if your relationship is ready for a child.

Few potential mothers really appreciate how much time and energy they need to devote to a child. This starts right from pregnancy onwards with doctor's appointments and planning for the birth. Preparing for a child takes time, energy, money and freedom.

It is unfair to bring children into the world if you haven't got the time and energy to devote to raising them. If you have a very demanding career you may not be able to devote yourself to a child. It is important to look at how your life is structured – what you are concentrating your energies on – to see if it is conducive to having a child. Would you be too preoccupied with your other interests to have a child? Or are you prepared to rearrange your lifestyle around a child?

Some lifestyles are more suitable to child-rearing than others. Households where both partners work long hours and don't have the time, households where there is fighting, drinking, drug-taking or gambling, i.e. engaging in activities which risk the child's security, are obviously not suitable. A child needs stability, security and consistency. You need to ask yourself if you can give your child that.

If you still have a lot of goals in life and you don't feel that you want to compromise on them for the sake of a child, perhaps now is not a good time to have a baby. Success, love, social life, security, travel, creative endeavours, education, may be more important goals to you at the moment. If this is the case, give yourself time to complete them before you decide to have a baby. If time is running out, decide if you are prepared to make a baby the priority in your life right now.

Not often mentioned but a major concern of the potential parent is the expense of rearing a child. Even daily child maintenance can be costly. Food, clothing, nappies, childcare costs – the list seems endless. In addition to everyday expenses there is money for education, entertainment, emergencies, illness, insurance and perhaps college. When making the baby decision a woman needs to assess her current income and future earnings. It might be helpful to put together a mock budget so that a realistic assessment can be made.

## Couples who can't agree about whether or not to have children

You want a baby and he doesn't; then a month later, he wants one and you don't. You just can't agree and so you argue.

There are ways to resolve the dilemma. First of all you must make sure you're totally clear in your own mind about what you want. If you are unsure that your partner is 'the one', talking about it is vital. You may want to consider counselling if you both feel you are getting nowhere. Fear of change or commitment, or of losing your partner, could be holding you back. You should discuss the practicalities – jobs, finances, childcare, housework and so on. If you are worried about fertility it might be worth getting his sperm checked and finding out if you are ovulating.

Should you decide to go ahead, eat healthily, cut down on alcohol, stop smoking and take a daily 400mg of folic acid (which helps prevent neural tube defects, such as spina bifida, in unborn babies). If you can't come to a decision, stop talking about babies and get on with your life. Set a time and date when you will have your next discussion about whether or not your relationship is ready for a child.

## Is your relationship ready for a baby?

If you are in a relationship, to be fair to the child both of you need to be ready. It is important that no one is pressured into having children against their will. Should the time come when you really feel you can't postpone any longer, you will have to decide what is more important to you, the baby or the relationship.

For a relationship to be ready it must be able to withstand the stresses and pressures a third person will bring to it. Most couples know that a baby will bring changes to their lives, but few really understand quite how much they will change.

When a baby arrives on the scene the present relationship you have will change. Your priorities and needs diverge dramatically when you become parents. It is helpful to know what these changes might be before you decide whether to become parents. In their book *The Transition to Parenthood: How a First Child Changes a Marriage*, Jay Belsky and John Kelly (Delacorte, New York, 1994) outline research done by the Penn State Child and Family Development into how a baby affects a marriage.

Their conclusion was that men and women experience the transition to parenthood in very different ways. Most of the transition occurs for the woman since it is her life that changes most dramatically. Much of the upheaval, such as the love and joy a baby brings, is positive, but there is also fatigue, anxiety, low self-esteem, loss of identity and mood swings. For men, although there is a certain amount of upheaval, the transition is easier to control. Stress levels tend to be lower than those of the mother.

According to the authors of *The Transition to Parenthood*, there are five areas that strain the relationship the most: division of labour, money, work, their relationship and social life.

Division of chores tends to be the biggest stress factor. Compared to many decades ago men today are far more actively involved in childcare routines, but new mothers still feel they bear the main burden, even when they work full time. Many new mothers feel that they don't get the help and the support they need from their partners. And fathers feel that they don't get the appreciation they are entitled to when they do fully participate in childcare and that their partner's criticism of their parenting skills is demotivating.

Couples contemplating parenthood would be advised to discuss the issue of division of labour and who is responsible for what before the baby arrives.

Money is another big stressor. Mothers tend to feel that no expense must be spared when baby arrives. Fathers, on the other hand, often want to cut back on consumption and increase financial resources. The result is conflict. It is especially hard if a woman has previously been working and making financial decisions for the household and now has to be dependent on the husband's financial decision-making. If, prior to the baby being born, a couple can work out a financial strategy where both feel they contribute, this can relieve a lot of tension.

Loss of libido is another huge adjustment in a relationship. A recent University of Michigan study found that the incidence of sex stops by 40 per cent in the first year after a baby. Fatigue is the most obvious reason. Focus on the new baby rather than the relationship is another.

Many new parents feel less emotionally connected to each other when they have a baby. Men often feel that their wives give the baby all the attention and women tend to think that their partners are being selfish for wanting their attention. Anticipating how a baby is going to affect your relationship and the quality and quantity of sexual intercourse can help avoid potential problems. Communication is vital.

Another source of divisiveness is that women expect their partners to share equal emotional responsibility for the child and to make career sacrifices. But although men today are willing to change the nappies and get more involved, they still tend to be emotionally unprepared for full parenting responsibility. Taking care of the child when it is ill, arranging day-care etc., are still regarded as women's work. Men also still tend to view themselves as the main breadwinner and are unwilling to make career sacrifices. If a couple can address some of their values regarding child-rearing, and how much sacrifice each is prepared to make, before the baby is born, this will ease the transition.

Studies show that social activity also declines when a baby arrives. New mothers tend to suffer more from isolation than new fathers, but both often find the lack of social stimulation frustrating. Fathers often blame their partners for being obsessed with the baby and mothers often feel their husbands don't understand how much they need to be with their babies.

The *Transition* study states that a woman's life tends to change more than a man's, but more recent research is proving that a staggering number of men do go through an identity crisis after the baby is born. According to Dr Malcolm George (quoted by Jane Phillmore in *Red* magazine feature, 'Boys Don't Cry', April 2000, p. 37), the discrepancy between the expectation of events and the reality of them can result in male post-natal depression. We hear little about male post-baby depression because men tend to hide it well.

How much fatherhood changes and alters a man's life is still a neglected area of research, but one thing is clear. A baby changes the lives of both father and mother and the nature of their relationship. Before a couple decide to have a child it is important both are aware of potential areas of conflict and disharmony and how the pressures of having a child are going to impact their lives and their relationship.

According to the *Transition* study, certain characteristics seem to indicate if a relationship is ready for a child. Can the couple:

- discuss and resolve differences about division of labour?
- work together as a team and surrender individual goals?
- handle stress effectively?
- argue constructively and maintain common interests?
- appreciate that the relationship will change and things cannot go back to the way they were?
- keep the lines of communication open?

No relationship is perfect, and a certain amount of stress and conflict is inevitable, but if mutual empathy and understanding can be achieved before the baby arrives this is a positive sign that a couple is ready to have a child.

## Coping with ambivalence

> 'I think I would feel terrible if I didn't have a baby, but I just don't know if I can make all the necessary sacrifices.' (Linda, age thirty-four)

> 'I'd feel selfish if I didn't have a baby, but my career doesn't leave me much time for a child.' (Samantha, age thirty-six)

If you are contemplating motherhood, expect to feel torn between conflicting emotions. Motherhood is a tremendous responsibility. Even today in our liberated society most women end up being the primary care-taker. Many women think they want babies, but the responsibility is frightening. They just don't know if they are willing to devote themselves to someone else for the next eighteen years.

Other reasons for ambivalence include fears of transforming into your own mother. Women who decide not to have children are choosing not to follow their mother's example in an important way. There could also be a fear of pregnancy and the bodily changes it brings. Of not losing the weight afterwards. Of becoming sexually unattractive as mother. Many women fear that a child will rob them of their independence or a hard-won career. If a woman is single or in a gay relationship she may have her reservations about raising a child in an unconventional way. Other women feel that the whole baby thing is far too grown-up. They haven't finished being children themselves. If you were inadequately parented, you may feel that you've got your hands full just finishing your own upbringing.

However much you want children, there are bound to be times when you feel overwhelmed by the thought of them. Every potential mother experiences ambivalent feelings. Hesitation before such an important step is perfectly natural. Ambivalence about parenthood is important for the decision-making process and should be viewed in a positive light. It shows how seriously you are taking the decision, how realistic you are being about potential parenthood and how it might affect you.

The question is, can you differentiate between natural ambivalence and the hesitation that deserves to be taken very seriously indeed?

Are your fears simply signs that there are certain things you need to resolve in your life before you have a baby, or are they are clear warnings that you shouldn't become a mother?

## Who should be mother?

> Who should be mother? . . . Now that we no longer require women to produce as many babies as possible . . . women cannot avoid making a choice whether to realise their reproductive potential or not. Fewer and fewer seem to be choosing the motherhood option and we are constantly told that those who do are the wrong ones. (Germaine Greer, *The Whole Woman* (Doubleday, London, 1999), p. 202)

You may be concerned that you are one of the 'wrong ones'. You may feel that you were not parented well by your parents and that you might repeat the same mistakes. That you haven't got the right personality to be a good mother. You may be aware of certain shortcomings or personal problems which you think might make you a poor parent, such as a quick temper, lack of patience, selfishness, perfectionism and so on.

Vital for good parenting is the ability to bond and love and share intimacy with another human being. We gain this capacity by experiencing it from others. The relationship between infant and nurturer is the single most important factor contributing to the emotional and physical health of a child. Through the love of our mother and father we learn to love ourselves and others. Every potential mother should think about her capacity for loving. We tend to love as we have been loved, so looking closely at your childhood can bring forth a multitude of emotions. Thankfully most of us have had happy, positive childhoods, but there are also many who haven't. There may have been physical abuse, sexual abuse or emotional incest when a parent violates a child's boundaries. There may also have been emotional abuse which is any kind of abuse that is not physical in nature. It can be anything from verbal criticism to intimidation to subtle tactics. It is a form of brainwashing that wears down a child's self-confidence. Many people are not aware that they were emotionally abused as children, and grew up thinking that the way they were treated was normal, when in fact it was abusive. But the experience left them with severe emotional scars.

Just because a woman was abused as a child does not mean she will abuse her own child, but there is a higher risk of the pattern repeating itself and abuse occurring. If you were abused in any way it is important that you work on the attitudes, beliefs and behaviours that you grew up with, which can cause abusive parenting. Experts and counsellors can help you do this by looking at your personality characteristics and your family tree for signs of neglect and abuse and addictive behaviours like alcoholism. Many people who have been abused as children long for the opportunity to give their children the kind of love and support and healthy parenting they never had. They are able to work through the issues and don't pass on the abuse.

Our generation has more choice and information than our parents did. We live in an era when parents are not expected to know instinctively how to raise a child. Books, information, magazines, parent groups and telephone hotlines are all there for us to discuss our concerns. Today all kinds of help are readily available from organisations which teach parenting skills and educate potential parents.

You can actually learn to be a good parent. The Parent Network (Tel. 0808 800 2222), The Parent Company (Tel. 020 7935 9635) and Parentalk (Tel. 0700 200 0500) all run courses which teach parenting skills.

If you are worried about your mothering skills, points to consider include:

- You have to accept that children won't love or want you all the time.
- You will have to love your child whatever he or she is like.
- Understand that if you respect your child, they will respect you.
- You need to teach a child boundaries and provide proper discipline and limits.
- You need to feel comfortable showing physical affection.
- Remember that being a parent is a work in progress. You will make mistakes.

Every potential mother wonders if she will be a good one. Feeling apprehensive and lacking confidence is not uncommon, but some personality traits are not conducive to good parenting.

Sensitive areas include how well you can control your emotions, how you handle stress, how patient you can be, how disciplined and consistent you are, how much self-esteem you have and how well you can give of yourself to others. There is no such thing as an ideal parent personality but patience, tolerance, flexibility, self-sacrifice and tolerance of intrusion are key elements in good parenting.

Parenting requires an incredible degree of patience. Children can be chaotic, unpredictable, stubborn, silly and rarely co-operative. You can't schedule or organise them to a timetable. Some of us are more impatient than others, but without patience you won't survive as a parent unless you become abusive or dictatorial, which is not positive for you or for your child.

Flexibility goes hand in hand with having patience. Since children are unpredictable, plans will have to be adapted at a moment's notice. You need to have the flexibility to accept what your children choose to do or not to do and to know which battles are worth fighting and which are not.

Having a child also means losing your personal space. Children are intrusive and demanding of your attention. In their eyes you and the rest of the world exist for them. Children love noise, lights, colour and constant activity. If you are very sensitive to intrusion and require your privacy and time alone you need to think carefully about how you will handle the stress of parenting.

Children can be emotionally draining. They need you all the time. You can't ignore them. You have to be able to put your needs on hold. If you think that this might cause you to feel resentment or to hate your child, or if you have problems putting your own needs aside because they are too great, you need to consider carefully if you are suited for motherhood.

Thinking about your attitude towards children in general is also helpful. Do you actually like children? Do you enjoy spending time with them? Many women who say they want children have had very little exposure to them. They may not even feel comfortable around them.

If you know you are not patient, tolerant, flexible and comfortable around children, this does not mean that you won't make a good parent. It just means that you are thinking about your personality clearly and realistically, and that you are aware of the areas you need to work on and the changes that need to be made in yourself.

Making the decision about babies is difficult. It requires a willingness to think not just what is best for you but what is best for you, your partner, if you have one, and for your child. The process won't be easy, particularly if there is the added pressure from the biological clock, but careful, honest self-assessment is the only way you are ever going to make a decision that is right for you.

And while you are thinking through all these issues, how much time do you have left before the biological clock strikes midnight?

# 6

# Can I Beat the Clock?

*The cosmic joke today is that you spend the first half of your life trying not to get pregnant: then you spend the second half trying to get knocked up.*
(Gail Sheehy, *New Passages* (HarperCollins, London, 1996), p. 109)

Can you actually beat the biological clock? What are the facts today about fertility and age? What options are open to you? How long can you postpone the baby decision?

## When the clock winds down

There's no denying the clock. You can convince yourself that there's still plenty of time to make it in your career, meet the man or woman of your dreams, climb Everest, whatever. But your body is telling you that it's not going to wait very much longer for that other big-ticket item on your To Do in Life list: Babies. (Julie Tilsner, *29 and Counting* (Contemporary books, Illinois, 1998), p. 63)

Fertility experts agree that a woman begins to ovulate less frequently from her early thirties and there is considerable evidence to suggest that her eggs get less fertile with age. Biologically speaking, the younger you are when you try to get pregnant the easier it is.

Conception will take longer when a woman is older. The 'average couple' takes six months to a year to conceive. But after the age of thirty getting pregnant can take up to two years. And during that time the body does not stand still – fertility continues to decline.

A girl is born with exactly the number of eggs she will ever have. So by the age of forty not only will there be fewer eggs, but the ones remaining will have undergone some deterioration. There's a drastic difference between the egg of a twenty-year-old and the egg of a forty-year-old. An older woman's eggs tend to be more irregular shaped, dark and granular. The outer cell layer is thickened into a tight mesh that makes it harder for sperm to penetrate. Older eggs may never be healthy enough for conception to occur, and if it does they may miscarry because of their relatively disordered condition.

There is also speculation that a number of systemic diseases, some drugs and possibly environmental factors, such as radiation and chemicals, can negatively affect fertility. The longer a woman lives, the more exposure she will have to their effects. It is believed that toxins, like smoking, factory pollutants, car fumes, pesticides and so on, can put more of a strain on fertility than that well-known blame all for fertility problems – stress. Emotional stress is probably more reversible than environmental toxins, which can play a major role in thickening women's tissues, creating so-called unexplained infertility.

The longer a woman leaves childbearing, the more likely it is also that fertility will be compromised by baby-unfriendly conditions, like infection or endometriosis. The infection chlamydia is symptom-free, but can cause scarring of the fallopian tubes. Endometriosis, far more likely in older, childless women because they have had more periods, is a condition in which menstrual blood backs up and adheres to the internal organs and bleeds each month. It can damage internal organs, such as the ovaries and tubes, and cause infertility.

All these factors begin to look rather negative for a woman delaying the baby decision. But it is important to point out that for every woman who has problems conceiving in her late thirties or early forties there is another who is astonishingly fertile until an advanced age. Cherie Blair, the British Prime Minister's wife, who gave birth at age forty-five, springs to mind. However fertile she is, though, a time will come to every woman when she can't conceive any more. It is just that the path each woman travels to that point is unique to her physically and emotionally.

The biological clock is inscrutable. That's what makes the question of when a woman's fertility will end, when it is no longer possible for her to have babies, difficult to answer. With all this uncertainty about potential fertility, it is not surprising that we get anxious about the winding down of our biological clock. It is very stressful having to make hard-and-fast decisions about the rest of our lives when the most important factor which will affect our decision remains elusive.

But even though we can't say with any certainty when a woman's fertility will actually end, there is one thing we do know, which many women don't fully appreciate. The process of fertility ending, known as perimenopause and menopause, does not happen overnight. It takes time, spreading itself over a number of years.

## Perimenopause

'Sounds cute, doesn't it, the peri-meni, but don't be fooled. This five to ten years pre-shutdown is when you feel most loopy' (Mary Spillane and Victoria McKee, *Ultra Age: Every Woman's Guide to Facing the Future* (Macmillan, London, 1999), p. 160).

Perimenopause is the period of five to ten years before menstruation and fertility stop at menopause, when we experience declining, fluctuating sex hormone levels. It's impossible to be precise about the age a woman will reach perimenopause, but it usually occurs between the age of thirty-five and fifty. For some women, though, it happens much earlier.

How long perimenopause lasts and how a woman experiences it has a lot to do with hormones and genes, but also a lot to do with her lifestyle, nutrition, activity levels, stress levels and other factors. But even though lifestyle changes can ease symptoms, individual perimenopausal differences among women do not reflect their comparative states of health. You can be very healthy and still genetically predisposed to perimenopausal symptoms. Perimenopause is not an illness but a perfectly natural transition before the menopause.

Perimenopause involves disruption to the normal menstrual cycle and a decline in fertility. The process is different for every woman and can affect the whole body. Here, listed as comprehensibly as practicable, are possible symptoms of perimenopause:

> Skin problems, like acne, dry skin, wrinkles and age spots, dark circles under the eyes, headaches, migraines, nausea, dizziness, feeling irritable, swollen ankles and feet, low blood sugar, trembling, anxiety, aches and pains, bloating, blood sugar imbalance, bone loss, breast changes, depression, hair growth, fatigue, muddled thinking, hair loss or thinning, hot flushes, infertility, insomnia, leg cramps, low noise tolerance, irregular periods, markedly worse PMS, migraines, mood swings, muscular weakness, night sweats, panic attacks, loss of interest in sex, sleep disturbances, stomach cramps, teeth troubles, incontinence, vaginal dryness, urinary infections, water retention, weight gain and inability to lose it, weeping.

This is not to say that all these symptoms will occur. No two women are the same, and each will have a different experience of perimenopause. Some of us may get insomnia, weight gain and irregular periods, while others get the irregular periods and the mood swings but not the insomnia. Some of us may suffer greatly, others will barely notice any problems.

Symptoms of perimenopause can be bewildering. Sometimes it can be hard to recognise what exactly is going on as perimenopausal symptoms are often similar to symptoms of Pre-Menstrual Syndrome (PMS).

In her excellent manual on perimenopause, *Before the Change*, Louise Gittleman sheds some light on the issue and tells us how we can distinguish between symptoms of perimenopause and PMS: 'If your period continues to be regular it's PMS. If your periods are irregular its perimenopause' (*Before the Change: Taking Charge of Your Perimenopause* (HarperSanFrancisco, 1998, p. 7).

Many women aren't really sure what an irregular cycle is. Generally a woman gets her periods anywhere from twenty-one to thirty-five days apart and bleeding lasts around five days. Lifestyle changes, such as simple weight loss, a stress overload, travel and so on can throw off an otherwise normal cycle causing a woman to miss a period, have a late period or menstruate twice a month. This isn't unusual. But if there is persistent irregularity for more than three or four months, and a woman is in her thirties and forties, the chances are she is entering perimenopause.

At some point during perimenopause a woman stops ovulating regularly. She may skip an occasional period or go for two or more successive months without ovulating. If no egg is released during a menstrual cycle it is impossible to get pregnant. Since there is no way of knowing when a woman will stop ovulating altogether this can be a particularly difficult and stressful time to make decisions about babies.

Delaying one or two years while you are in perimenopause could prove crucial.

## Menopause

'Menopause is as individual as a thumbprint. No two women experience it alike' (Gail Sheehy, *New Passages*, p. 209).

Menopause follows perimenopause. During the menopause the ovaries make fewer and fewer hormones. Cycles become even more irregular and eventually they cease altogether. A woman's reproductive life is over, barring medical or divine intervention. Midnight has struck on the biological clock.

Symptoms that accompany menopause are similar to those of perimenopause. Again they vary from woman to woman. Some women suffer terribly during menopause, others only experience mild discomfort.

The great majority of women go into menopause in their late forties and early fifties, but it can happen much earlier. Some women only discover when they are trying to conceive that they have suffered a premature menopause. It is possible to stop menstruating any time from the twenties to early forties because of a premature failure of the ovaries.

'Since 7 to 11 per cent of women go through premature menopause before age forty, according to Dr Richard Bronson, physicians should not tell women they won't go through menopause until they are about 50' (Gillian Ford, *Listening to Your Hormones* (Prima, California, 1998), p. 44).

Menopause, like adolescence, is a perfectly natural transition in a woman's life. It should be an exciting one. According to Joan Borysenko, menopause is a

> second puberty, an initiation into what is the most powerful, exciting and fulfilling part of a woman's life. The time when we can find fulfilment both personally and socially. In freeing us from the reproductive cycle this stage in life offers us the opportunity to fully understand ourselves, our sexuality and what we have to contribute to the world. (Joan Borysenko, *A Woman's Book of Life* (Riverhead, New York, 1996), p. 140)

But if a woman still hasn't made up her mind about motherhood, menopause is not something to be celebrated. More likely, it is something she will fear.

## The reproductive prime

> 'The medical world has a term for a woman having her baby for the first time over the age of thirty-five – elderly primigravida – which says it all really.' (Jane, first-time mother at forty-two)

With perimenopause and menopause many years away, a woman is indisputably in her physical and reproductive prime in her twenties. Physiologically the twenties are the ideal time to get pregnant, because miscarriage, infertility and medical risks are at their lowest point. For example, a woman in her forties is thirteen times more likely to have a baby with Down's syndrome than a woman in her twenties, the National Centre for Health Statistics reports.

Yet the trend today in Britain and the USA is to postpone childbearing until the thirties when women are more likely to have settled into a career, a relationship and come to terms with who they are. Since 1980 the birth rate for women in their thirties has nearly doubled. One projection states that in a few years, roughly one in every twelve babies will be born to women thirty-five and older.

But getting pregnant isn't always so simple after the age of thirty. In her thirties a woman sees a gentle decline in fertility. A 5 to 10 per cent drop takes place between thirty and thirty-six and after that a much steeper decline. Women of forty are less than half as efficient at conceiving as at thirty. There's a higher chance of miscarriage and abnormality. By forty-five a woman is lucky if she conceives. The drop is attributed to lower ovulation rates and lower egg quality, which decreases the chance of embryo development and increases the risk of chromosomal abnormalities. And once menopause occurs and egg production stops altogether, getting pregnant is virtually impossible.

According to most medical authorities a woman's biological reproductive prime is in her twenties. But experts also agree that the best time for a woman to have a baby is not just when she is physically ready, but when she is emotionally ready as well. This makes the ideal time for motherhood as individual as DNA. A woman may be in her reproductive prime, but she may not be emotionally ready for a child. She could decide to wait but then have problems conceiving. Either way there are positives and negatives. In the words of Professor Helen Haste, author of *The Sexual Metaphor* (Harvard University Press, 1994),

> We all mature at different rates. The key thing about childbearing is that it's a major, but normal life crisis, which causes substantial

> change to one's life and identity as a person. Whenever you do it, it's a major upheaval. Ideally you plan it to fit your own life script.

And should a woman decide it is best for her to wait to have a baby, what are her options? Even though nature may be telling her that she could be leaving it too late, can she beat the clock?

## Miracles of science

In 1996 Dr Richard Paulson assisted 63-year-old Arceli Keh to conceive using in vitro fertilisation with a donor egg. Nine months later she became the oldest woman in the world to become a mother. This incredible example of how science and technology can change the course of nature has raised tough biological, emotional and ethical issues that will be debated for years to come.

More and more women are beating the odds and having healthy babies later. Even though it's harder to get pregnant after the age of thirty-seven, some women are more naturally fertile than others. The elderly primigravida is out there and her baby is thriving. And if a woman has postponed the baby decision and can't conceive the natural way, many options are open to her. The biological clock used to come to a dead halt when a woman's body couldn't perform the magic task of conception. But that's not the case any more.

We live in a world that is giving hope to thousands of women who have despaired of ever having a child. The extraordinary rapid advances in reproductive endocrinology are dizzying. Older women with declining fertility levels can take fertility drugs or treatments to help them conceive. If their eggs are faulty they can borrow them from someone else. It makes no difference today if a woman is gay or straight, single or married, ill or well, young or old – in one way or another new reproductive technology can bypass most of these barriers. The trend towards late motherhood has taken a dramatic and controversial step further.

With the use of donor eggs taken from the ovaries of younger women, it is perfectly possible to get any human female pregnant, be she seventy years old or merely seven. This may be the case, but most fertility doctors are very reluctant to treat women over forty-five. This isn't just because fertility takes a nosedive after thirty-nine and success rates are low, but because post-menopausal pregnancy is a sticky issue.

Our society isn't comfortable with the idea of wrinkles and babies going together. Some argue that midlife women had their chance and wasted it. They chose to build careers, travel and experiment with relationships, and now they have missed their biological deadline they want another chance. Others say that it is unnatural. Women were meant to be fertile for only a limited period of time and we should keep it that way.

Many argue that reproductive technology should only be used for women who lost their chances of conceiving through illness or surgery or premature menopause. It should not be used for older women who want babies. The whole issue of egg donation and who is the 'real' mother complicates things further.

Many of these concerns about older mothers are legitimate. But then again, isn't it every woman's absolute right to have a baby? Why can't a woman in her forties and fifties with the maturity to love and give of herself have a baby? If a woman got pregnant by accident in her forties or fifties nobody complains. But if you seek out treatment you are suddenly under the spotlight. Is this fair? As a society, is it really in our best interests to regulate who should and should not mother? You could argue that post-menopausal pregnancy is simply a new way of respecting individual choice. Every day our capabilities are shifting. Isn't it time that we started to rethink our attitudes about the stages of human life?

The people who make the best parents are those that are committed, motivated, loving and caring. Accounts of post-menopausal births speak enthusiastically of the love, time and commitment the mother can give her child. This is no doubt true, but serious questions do need to be raised when a woman wants to become a mother in her fifties and sixties. The most important question is what is in the best interests of the child? What does it mean for a girl of sixteen to have a 70-year-old mother? Who will she learn her role cues from? Who will she have a loving, playful, energetic relationship with?

And has a woman in her fifties and sixties enough energy and motivation to raise a child? Having a baby is tough enough in your thirties. Imagine what it would be like in your sixties. And what if the mother gets a terminal illness? What will happen to the child?

Sadly, in some instances, the longing for a child can be so great that it may surpass what is actually in the best interests of the child.

## Racing to beat the odds

Millions of older women have undergone fertility treatment and proved that they can have healthy babies. The problem is that now that it's technologically possible to put off having children, it's much easier to postpone the question until it really is too late.

Some women find that they have placed too much faith in the miracles that modern medical science can perform. They overlook the hazards of ignoring their biological imperatives and find themselves beaten by the harshest deadline of their lives: their own reproductive clocks.

> Emily had wanted to become a mother at age thirty. But her life as a busy teacher and writer made settling down difficult. She didn't meet the right partner until she was thirty-eight. She took her fertility for granted and got pregnant at thirty-nine after trying for two months. Her photograph album shows a grinning woman pointing to a positive pregnancy test. Nearly three months later she was carrying a dead fetus and had to abort the first of four babies-to-be.

> At age forty-four, Carolyn, after three months of a healthy pregnancy – with insatiable hunger, nausea, exhaustion, joy and three normal sonograms – had her fourth and final miscarriage. A lab report showed that the baby had been a boy with Down's

> syndrome. Carolyn describes how she and her husband had seen him on a sonogram three weeks earlier. He had the oversize head typical of a young fetus, and his tiny feet and hands were waving wildly from his little body. 'Miscarriages are awkward,' she says. 'Nobody knows what to say. They are so often trivialised – especially if they are in the first trimester. Everybody urges you to try again. The grief is unbearable and hospitals don't know how to deal with it. The sonogram nurse just let me watch my lifeless child on the screen as she left the room to get the doctor.'

Many women today assume that they have until menopause to have a baby. They may not. For every woman in her forties who becomes a first-time mother there is a woman who has waited until her late thirties or forties and cannot conceive. Sometimes even the most expensive, advanced fertility treatments are unsuccessful and a woman can't have her own baby.

Sadly, some women who postpone the baby decision until the last moment have been beguiled into believing that the decision to have a baby or not can be prolonged indefinitely. They may really have started to believe in the fantasy of 'fertility forever'. They may have left it too late to have a baby, but before such a limitation is accepted, there comes the desperate race to beat the reproductive clock.

An older woman may feel that she has reached a point in her life when she needs something more. She doesn't want to miss out on the nurturing experience. She has always believed that her choices wouldn't be limited by age, that medical science can extend her biological clock. When she discovers that she can't get pregnant, her quest for a baby may become all-consuming.

Several miscarriages may lead to fertility treatment. The treatment is expensive, but if the urge to have a baby is strong she could be prepared to make any sacrifice. She may choose to put her body through hell. High doses of hormones may be pumped through her system. If her own eggs are faulty she may consider using those of a donor. She often puts her life on hold. Starting over and over again each month gets harder and harder. But she is likely to persist until the last shred of hope has gone.

Frustration, anger, denial and depression are likely to set in as she watches that hope slowly evaporate.

> 'Time is running out and I'm really feeling wretched and desperate. Every part of me is crying out to have a child. I feel so cheated. So empty. So worthless. I've got the education, the qualifications, the career. I've got everything it takes and yet I can't have a baby. It doesn't make sense. I think of all those children born in poverty who have to suffer so much. All those teenage mothers with unwanted pregnancies.' (Sharon, age forty-one)

In addition to mourning the loss of a dream there is the death of the illusion of control over her life. The older woman has grown accustomed to autonomy. She probably practised birth control for many years. She has her identity and her life. Now infertility robs her of all that. Suddenly her body is controlling her. As the window of opportunity closes with age, failure to conceive can completely devastate her self-worth.

It really is a lottery when it comes to fertility in the late thirties and forties. There are no guarantees. Yes, some women can beat the biological clock naturally or with medical help. But it is also true that many can't. Techniques, success rates and availability of infertility treatments are poor in the UK in comparison to the US and the rest of Europe. A new study reveals how the NHS is letting down couples for infertility. 'The shame of Britain's poor deal for infertile couples', writes journalist Beezy Marsh (*Daily Mail*, 2 July 1999, p. 24). Long waiting lists for treatment for an older woman reduce her chances of conceiving. IVF cycles are limited to one or two for cost reasons. Success rates are higher privately, but private treatment is expensive.

The harsh truth is that some women do postpone the baby decision too long. Even the most advanced fertility treatments can't give them a baby that is genetically theirs. There is the risky possibility of using donor eggs, but for the majority the race against the reproductive clock has been lost.

Or has it?

## Fertility frozen

> 'I didn't feel the overwhelming maternal urge and desperation some single women of my age might feel. I've always feared being pressured into marriage simply because of my biological clock. That is why, two years ago, I decided to give myself the freedom that freezing my eggs would provide.' (Sarah, age thirty-four)

In the 1990s many women who were about to undergo chemotherapy or radiation therapy for cancer, and lose ovarian function, began to ask doctors if it might be possible to freeze their eggs. If they already had a partner they could freeze both egg and sperm as an embryo. But if a woman didn't have a partner she would need to save her eggs.

The success rates for freezing eggs were initially very low, and today freezing eggs is a lot less efficient than embryo freezing, but it is now nevertheless a viable option. In October 1997 a successful twin pregnancy resulting from a frozen-thawed egg donation was reported by Dr Michael Tucker, an embryologist at Reproductive Biology Associates in Atlanta. Egg-freezing is changing our concept of motherhood and the restrictions we put on it. If a young woman doesn't know if she wants children, worries that she may never meet the right man in time, doesn't want to be pressured by her biological clock, or has been told that she has a condition which prevents her from getting pregnant, she can opt to have her eggs frozen. Held in suspended animation at the temperature of liquid nitrogen these eggs can last for years, even decades, until she is ready to make use of them.

If a woman wants her eggs frozen she should get in touch with an assisted reproduction and gynaecology clinic prepared to undertake the procedure. In January 2000, women in the UK gained the right to use their own pre-frozen eggs to get pregnant, following the example of America, Japan, Australia and Italy. The legalisation of the procedure in several countries has opened the door to women who want to put off pregnancy till it suits them, as well as to women who face medical treatment that could damage their ovaries. It also makes it possible to set up frozen egg banks.

Costs vary, but tend to be around £2,500 in the UK and $8,000 in the US. The procedure takes thirty days. Freeze-thaw involves a trip to the clinic where the consultant will check if it's suitable for you. There will then be blood tests during the first couple of days of your menstrual cycle to check hormone levels and the right stimulation routine for you. When you are ready, you'll receive daily hormone treatment (often injections) to stimulate your ovaries; scans and blood tests monitor the maturing egg follicles. Once the follicles are the right size, you'll be booked in as an outpatient for egg collection under a general anaesthetic or intravenous sedation. The follicles grow on the ovaries. A probe is put in your vagina and under ultrasound guidance a very fine needle punctures through the vaginal wall into the ovary. The fluid from each follicle is aspirated down a tube, hopefully with a ripe egg. Up to thirty eggs may be collected. The whole process takes around thirty minutes.

The eggs are then cultured in the laboratory and prepared for freezing, usually by adding antifreeze and slowly dropping the temperature. They are then stored in liquid nitrogen at −196 °C. The biological clock stops. In theory they could be frozen for hundreds of years. In the UK there is a ten-year time limit.

Thawing for potential pregnancy is rapid. The antifreeze that protects the egg is gradually diluted to a normal culture medium of salts and proteins. After direct injection of a single sperm with a micro-needle, the fertilised egg can be implanted in the uterus. Within fourteen days you will know if you are pregnant.

Experience has come from countries like Italy, Australia and the US where there is no regulation. The success of freeze-thawing rose sharply since 1997 when researchers started to inject sperm directly into the thawed egg, using a technique called intracytoplasmic injection (ICSI). ICSI overcomes problems caused by the toughening of the outer membrane of the egg during the freeze-thaw process.

In the UK two clinics are allowed to freeze eggs: ARGC in London, and the Centre for Assisted Reproduction (CARE) at the Park Hospital, Nottingham which has had a licence to freeze since December 1999. Only ARGC has a formal licence to thaw eggs. Currently the Human Fertilization and Embryology Authority (HFEA) is considering applications to thaw eggs from other clinics. There is nothing to stop British women freezing and storing their eggs abroad. One woman from the Lake District became pregnant from freeze-thaw eggs at the Florida Institute of Reproductive Medicine (FIRM) which works in collaboration with the ARGC in London.

In theory, if a woman decides in her forties or fifties to have a baby and can't conceive naturally she can retrieve her healthy, young eggs for fertility treatment. In reality, though, the technique of freezing eggs is still in its infancy.

Sperm has been successfully freeze-thawed for the past forty years but eggs are far less easy to freeze because of their size. They contain a lot of water which can form ice crystals that could damage the egg. Their chromosomes also tend to scatter on freezing so, when the egg divides, the two halves won't necessarily contain an equal number of chromosomes.

There is still considerable debate about how long a woman's eggs can safely be frozen and whether or not the procedure will damage them. The big difference is also that if you make a mistake with sperm there are millions more!

Between 1986 (when the first freeze-thaw baby was born) and 1997 only three births were reported worldwide. But since 1997 results have been more encouraging and twenty-five births of healthy babies have followed. However, freeze-thawing eggs is still an uncertain, experimental technique.

Any woman contemplating egg-freezing must be informed of the risks involved, the low success rate (the HFEA puts the success rate at no more than 1 to 10 per cent) and how invasive, expensive and uncertain the procedure is. If she is determined to go ahead she should make sure that she chooses a centre with good IVF techniques which help to optimise the condition of the oocytes (eggs) retrieved, where good results have been achieved with frozen embryos and where reasonable preliminary data on oocyte-freezing is available.

Even if a woman decides to have her eggs frozen this is no guarantee that she will be able to conceive a healthy baby. It will be a while before the procedure becomes more available and commonplace, especially in the UK. Until thousands of babies have been born healthy and followed up for years, no one can be certain how safe it is. Invasive antenatal tests which carry a risk of miscarriage, such as amniocentesis, are recommended should pregnancy occur.

So at present it looks like the biological clock may still be winning the race. But then again it might not. New developments in fertility treatment are taking place all the time.

One of the most recent advances in fertility treatment which promises to be more successful than egg-freezing is ovarian-tissue freezing. It has been found that microscopic pieces of ovarian tissue can be frozen, thawed and then cultured to produce eggs. The benefits of taking a slice of ovarian tissue is that it will potentially have hundreds of follicles and each could produce an egg. The tissue can be frozen without causing any damage to the follicles. This enables a woman to store a far larger quantity of eggs, which increases the chances of a viable pregnancy. Instead of just freezing a few eggs the entire process that produces eggs would be frozen and it could be frozen at a time when a woman was young and healthy, limiting the chance that any future child would be born with defects. The tissue contains genetic material but is not usable in the state it is in. The immature follicles would have to be matured in a laboratory by an injection of hormones provided by a surgeon.

This technique has not been perfected yet, and it may take five to ten years to do so, but if successful, it could revolutionise fertility treatments in the future, because it bypasses the need for drug treatment to stimulate ovulation and enables a woman to store not just a handful but hundreds of potential eggs for use when she is older.

Who knows, in the future we may all have the option of storing our eggs so that we can conceive later in life. Maybe men and women will one day have the opportunity to save their sperm and their eggs for use later before making themselves infertile as the ideal form of contraception. We may all be given the chance to choose the right time for us without being dictated to by nature. Biological-clock anxiety could become a thing of the past.

And the fast-growing trend for women to postpone decisions about babies and have them later in life, discussed in the next chapter, could become the norm.

# 7

# Is Later Better?

'Motherhood going out of fashion for young women.'
'Single, over 35 and ready for Motherhood.'
'Baby boom time for Britain's Forty Somethings.'
'A trend is born as women wait until their 40s for a baby.'

The news of Cherie Blair's unexpected pregnancy at age forty-five inspired a number of news articles which confirmed what most of us knew already. There is a definite trend towards midlife parenthood. Statistics show that high-profile women like Cherie Blair, Emma Thompson, Madonna and Patricia Hodge, who have started families or had another child in their forties, are part of a sustained social change.

You can spot evidence of this trend at almost every PTA meeting, football match or ballet lesson. In the words of Irwin Matus PhD, a Denver psychologist and author of *Wrestling with Parenthood: Contemporary Dilemmas* (Gylantic, 1995), 'You can always see one or two parents with gray hair at these type of events . . . All the other parents are still dealing with their acne' (quoted from 'Beating the Biological Clock With Zest', American Psychological Association webpage – http://www.apa.org/monitor/feb96/fam40a.html, p. 1).

At the dawn of the new millennium the average age for first-time mothers is getting later and later. Currently it is hovering close to thirty years. Births to women in their twenties are declining and births to women in their thirties and forties steadily rising. More and more of us are weighing up the odds and deciding to mother later.

Since we live in an age of birth control, we have to assume that when a woman has a baby later in life, more often than not it is through choice. But what are the biological consequences of deferring? How does a woman's age affect her emotions and actions as a mother? What are the implications of the trend towards later motherhood for our society?

## Biological consequences of late childbearing

If you are undecided, you may be concerned about how much time you have left on your reproductive time clock, and the biological consequences of late childbearing. You may have a vague feeling that deferring the decision past the mid thirties increases the likelihood of problems, but you may not be aware of all the facts.

As discussed in the previous chapter, the first potential problem is fertility: having trouble getting a pregnancy started and avoiding miscarriage should conception occur.

The next hurdle is the higher risk of having a baby born with birth defects. Down's syndrome is the biggest fear that haunts the older woman. Down's syndrome is the single most common cause of mental retardation in the UK. It is not hereditary, but it is caused by the presence of an extra chromosome in the fetus. The disorder is clearly associated with maternal age. For a woman in her twenties the risk of having a Down's baby is low – about 1 in 1,500. In her early thirties the risk is about 1 in 750. In her late thirties the risk is 1 in 380. After forty the risk is 1 in 100, and after forty-five it is 1 in 40.

There are other major handicaps for the older mother to fear too: congenital malformations, such as heart defects or limb deformities, problems with the central nervous system, like spina bifida or inherited genetic diseases like cystic fibrosis and other chromosome disorders, like Down's, caused by poor separation of chromosomes at conception.

Fortunately more information is available to the older mother now than ever before through screening, which tells her the probability or risk of having a child born with a handicap, and diagnostic screening which can give a definite answer. Blood tests and ultrasound screening can also tell a lot about the health of a baby.

Every pregnant woman is confronted during her pregnancy by a series of options for antenatal testing, but women over the age of thirty-five are routinely advised to have an additional form of diagnostic testing called amniocentesis. In this procedure, performed around week sixteen, a sample of amniotic fluid is withdrawn from a pregnant woman's uterus. The fetal cells are then cultured and analysed. By examining the cells under a microscope technicians can spot abnormalities in the chromosomal pattern. If an abnormality is detected, the woman can decide if she wants to keep the baby or not. Another method of obtaining cells is called chorionic villis sampling, performed at around week fourteen.

Unlike blood tests, diagnostic testing is not always risk-free. Even when performed by a highly trained physician, amniocentesis carries with it a risk to the mother and baby. About 1 in 150 miscarry as a result.

Every pregnant woman worries about testing, but the older mother has more cause to. Chromosomal disorders, like Down's, rise with age because older eggs tend to be in a more deteriorated condition. Older eggs also increase the risk of miscarriage. So even if a woman is ovulating and gets pregnant she could be in a subfertile state. As a result the issues surrounding testing are always more pressured than those for the younger woman. The older woman is faced with a double-edged sword. She knows that her age is putting her at increased risk. She knows that a diagnostic test could be putting her longed-for baby at possible risk. She knows that if she miscarries, the ticking of the biological clock could take her to the wrong side of fertility.

Another age-related obstacle the older mother has to face is that she is immediately placed in the high-risk category by the medical profession. Doctors call a woman over thirty-five who is pregnant an 'elderly primipara' and treat her with a little more caution than a younger woman. This is because as middle age approaches, a woman is more likely to develop medical problems during pregnancy such as diabetes, hypertension and kidney and heart diseases, which can interfere with maternal–fetal interaction and jeopardise the health of both mother and baby.

According to Maggie Jones, author of *Motherhood After 35* (Fisher, Arizona, 1998), the risk of complications does increase during pregnancy and labour. Specific risks also include pre-eclampsia, haemorrhage and dysfunctional labour. Cervical dilation may be slower in older women. There could be less muscle tone and weaker contractions. Uterine fibroids tend to develop and grow with age and these can interfere with labour. Older mothers often have longer labours, but the difference isn't all that significant. There is a rise in obstetric complications. This may be due to an increased risk of placental failure for the older mother if the baby is overdue, so labour is more likely to be induced. The higher rate of Caesarean births may simply be caution on the part of doctors.

Not all doctors, however, believe that maternal age places a woman or her baby at greater risk. With prenatal tests and a better understanding of a woman's physiology, they feel that should problems arise they can deal with them. In fact, some firmly believe that the medical risks have been exaggerated, that the older woman can look forward to a normal pregnancy and a healthy baby and may even be at an advantage.

Dr Christiane Northrup, author of *Women's Bodies, Women's Wisdom* (Bantam, New York, 1998), has said that she would rather take care of a 40-year-old who looks after her health than a 25-year-old who smokes and eats junk food. Because older women have established themselves at work, they often have more time to enjoy their babies. They know what the world is like and are more willing to reassess their priorities. They are often more comfortable with their bodies than women in their twenties. For Dr Northrup, the older mother is low risk.

## Beating the biological clock with zest

As midlife mothers become more common, not only is the medical high-risk stereotype being challenged by some doctors, but other myths are breaking down too. She is no longer regarded as a narcissistic woman who squeezes her children in after work, says Diane Ehrensaft PhD, a psychology professor at Berkeley's Wright Institute. 'These women are actually satisfied and settled in their work. If anything they are overly zealous about being good mothers' (quoted from 'Beating the Biological Clock with Zest', American Psychological Association webpage – http://www.apa.org/monitor/feb96/fam40a.html, p. 1).

In fact, many psychologists now speculate that the very traits that characterise successful career women – such as effective problem-solving, organisational appetite and good communications skills – bode well for good mothering. 'Older women who decide to become mothers do so with the same zeal and ambition with which they've undertaken other things in their lives,' explains Rennee A. Cohen PhD, a psychologist in private practice in West Los Angeles. 'They are women who are used to postponing gratification' (quoted from 'Beating the Biological Clock with Zest', American Psychological Association webpage – http://www.apa.org/monitor/feb96/fam40a.html, p. 1).

Cohen also thinks that whether they continue working or not, older mums see a shift in their priorities and 'tend to have more balance in their lives'. She believes they have the patience and wisdom to cope better with frustrations and problems that may overwhelm a younger woman. If they have careers, they are used to negotiating and explaining things. These skills can help them set limits and convince their children that certain behaviours are not acceptable. They are also more likely to have done their reading and to know what to expect.

This all begins to sound very positive, but it is important to make clear that midlife motherhood is not always such an easy transition. Not all midlife mothers are as well adjusted as enthusiastic psychologists like Cohen purport.

> 'Ignore all that stuff you read about maturity,' says Carolyn, age forty-two. 'Unless you are prepared for children psychologically as well as physically, you're as out of your depth and as vulnerable as a teenager.'

Not all mature mums cope well with motherhood. Not all of them enjoy motherhood. Not all of them have thought carefully about becoming mothers. Some may have had a baby because they feared their fertile years were drawing to a close, or because they wanted to trap a mate, or to avoid loneliness in old age.

Generally, though, when a woman has a baby late in life it tends to be after much thought and effort. Most babies are wanted, but late babies – usually planned and longed for – are especially wanted. The older mother has beaten the odds and already turned convention on its head. She knows that by postponing motherhood she took a gamble.

Yet, however determined she may be to continue beating the biological clock with zest, she has to accept that late motherhood, although it will bring with it particular gratifications, will bring with it particular frustrations too.

## Frustrations and gratifications of late motherhood

From my interviews with older mums one thing became abundantly clear. They all agreed that there were both disadvantages and advantages to the choice they had made.

Some of the disadvantages older mothers presented included concerns about their own ageing and the possibility that they might not watch their children graduate, get married or have children. But many were able to turn this disadvantage into an advantage. It made them much more aware of the present moment. It also gave them a better perspective on time. However difficult they were finding the early years, maturity had taught them that nothing in life lasts for ever. Right now their child was taking up all their time, but they knew that soon their child would be at school and life would change again.

Another age-related concern was not having enough energy to 'keep up'. One area in which older mothers definitely lose out is that they do have less physical resilience. Research has shown that older women are less able to cope with sleep shortages in particular. Actress Patricia Hodge, who had sons at forty-two and forty-five, admitted that there were times when she wept from one end of the weekend to the next with sheer exhaustion.

The lack of verbal feedback in a child's early years could be frustrating for the maturer mother who often needs more stimulation than the infant can provide. Problems often centred around the older mother feeling that she wasn't stimulating or playing with her child enough. Several older mothers said they would probably have found the 'playing' aspect of motherhood easier if they had been younger.

> 'I'm not really a good playmate for my baby,' says Sarah, age thirty-nine. 'I just can't get down on my knees and crawl about playing peekaboo. It's not my thing. I don't enjoy it and I don't do it very well.'

Many older mothers felt that age brought greater maturity and an ability to handle change and crisis, but some said that they found the intrusion of a child on a previously established routine a difficult aspect of motherhood. They felt that having babies younger might have made them more spontaneous in their attitudes and less goal-orientated. They admitted that it was hard for them now to go with the flow. Some felt that if they had become mothers younger they would have had the advantage of growing up with their children so that their children could see that life was not always carefully planned and ordered – that it had its ups and downs. But the majority felt that being able to offer their children stability was more important.

Most believed that postponing babies was the only way they could have established themselves in a career. They were pessimistic about the idea of returning to work after having children young, believing that there just were not the opportunities for the returning thirty-something mother in the workplace.

Whatever age a woman chooses to mother she will face one of the most challenging, developmental tasks of her life. Motherhood is one way a woman can be pushed to work more intensely on her current life tasks and the areas of her life she may not feel complete in. 'Whatever the timing,' writes Carolyn Walter, 'parenthood is a mover and a shaker' (*The Timing of Motherhood* (Lexington, London, 1986), p. 106). It is a developmental, reciprocal process that engages a woman in caring for another human being and can facilitate growth in a less-developed side of her personality.

Younger women often say that the responsibilities of child-rearing builds their self-esteem. The adult tasks of separation and individuation are being achieved through motherhood. Older women find that having babies later involves moving away from already achieved autonomy and separation to a more caring, nurturing role.

From my interviews, what emerged as one of the biggest struggles for the older mother was the difficult balancing act between her own needs and those of the child.

> 'It's the loss of freedom, independence and choice I find the hardest to deal with. All those things I was used to for half my life are gone.' (Barbara, age forty-six)

The self–other dilemma is perhaps one of the greatest challenges of motherhood. It is particularly difficult for older mothers. How well they managed to balance their own emotional demands and those of the child contributed significantly to the degree of satisfaction they experienced in their relationships with their children.

The first year of motherhood seemed to be particularly daunting for older mothers. This is the year when the child is at its most dependent and for a woman with a strong sense of self there could be vulnerabilities regarding merging or closeness. Motherhood represents a very different lifestyle from the experience of working outside the home and older mothers are often forced to examine who they are in a very self-critical way. Many spoke of a loss of self-confidence in the early years of their child's life.

> 'At work I felt in control. I was a confident leader and manager. Alone with my baby I felt sadly inadequate, bewildered and desperately lonely.' (Linda, age thirty-seven)

Before childbearing older mothers have already established a firm sense of who they are. This can lead to a loss of self-confidence but it also means that they don't tend to view their children as extensions of themselves. They don't get lost so easily in the merging of mother and child. 'For late-timing mothers parenthood is a smaller part of their self-definition,' writes Carolyn Walter (*The Timing of Motherhood*, p. 22). Walter observes that older mothers seem to understand better than younger mothers that intense closeness with a child does not require that they give up who they are. Late-timing mothers enjoy the warmth and closeness of their roles as mothers, but are able to separate themselves and to perceive themselves in other roles. Walter goes on to say that whereas younger mothers tend to find discipline and separating from their child frustrating, older mothers may find that their greatest tension is lack of free time to pursue their own interests and meet some of their own emotional needs.

Most of the older mothers I spoke to agreed that the most difficult aspect of motherhood for them was not having enough time for themselves and for their partners. Those that worked reported a great deal of frustration with balancing their career and family: how difficult it was to plan a daily schedule; how their patience and tolerance grew short when something unexpected happened like illness, lack of sleep and so on. Frustrations tended to centre around loss of freedom.

Although the older mums I spoke to admitted that having babies later definitely had its disadvantages, most of them seemed satisfied with the choices they had made. Generally the consensus was that, although being younger would have had its advantages, it wouldn't necessarily have made them a better mother. Growing up before we have children seems to work as well as growing up with and through them.

This shouldn't be taken to mean, however, that later is necessarily better. Once again, it all depends on the individual woman. Women in the different decades of life can be equally competent in their roles as mothers. Each decade brings specific advantages and disadvantages. There is no 'perfect' age to mother.

Women of all ages describe the process of becoming responsible for another human being as one that weighs heavily with them but also fills them with joy. It will certainly have an incredible impact on your life, whatever your age.

> In the end I think that's what so great about babies. They throw your life in the air and, on balance, all the bits you can do without disappear and the good bits fall into place. Even sleep – eventually. (Quoted from an interview by Frankie McGowan, 'Have a first baby after 40', *Aura* magazine, May 2000, p. 109)

## Modern mothers

Some people believe that women are getting it all terribly wrong. They argue that modern life is being 'turned upside down' because so many of us are choosing to delay childbearing. Furthermore, they say, the biggest losers are women themselves.

*What Our Mothers Didn't Tell Us*, with its daunting subtitle *Why Happiness Eludes the Modern Woman* (Simon & Schuster, New York, 1999), argues that we have created a wasteland of desperate single, harassed mothers juggling work and home, and lonely, latchkey children. Plus overworked fertility experts trying to rewind the biological clock. The author of the book, Danielle Crittendon, is in her mid thirties and had her children young, then set about building a career.

The crux of Crittendon's argument is that she is trying to look at it from the ambitious woman's point of view. If you want to 'have it all', she advises you have the babies first and then devote yourself to career later. You won't be impeded by maternity leave and childcare issues. There will be no guilt about leaving your young children in someone else's care. And what's more, you won't end up anxious and sad in your thirties and forties desperately trying to find a partner and beat the biological clock.

Crittendon says that our generation of women had a sense of indefinite postponement in our twenties. We thought we would be fertile for ever. We felt omnipotent while we climbed the career ladder. Then we came down to earth with a bump when we realised fifteen years on that our fertility might be compromised and we might not be able to 'have it all'.

'Having it all' has become a catchphrase for women of our generation. We want the education, the career and the family. We want every option to be available to us. But while it is true that our life choices are expanding and changing rapidly, what has not changed much is provision for childcare and respect for the mother at home and in the workplace. Crittendon fails to take into account this rigidity in the workplace, and consequently her argument is one-sided. Because having babies early worked for her, she thinks it will work for everyone. But given the lack of cultural and corporate support for mothers and children, it is hardly surprising that more and more women are thinking twice about children. Women's choices are determined by the values and structures of the society they live in, and our society does not set its values or order its institutions in a way that suits mothers.

Young women are learning at a younger and younger age the harsher lessons of life. There may not always be a partner for them to lean on. Bringing up a child is expensive and exhausting. Having a child in your twenties can put limitations on career development. Building a career and raising a child is a frustrating juggling act. Women are taking these maxims to heart, weighing up the odds and, if they decide to have babies at all, having them later. It is the uncertainty of employment, the high cost of raising a child, the isolation and exhaustion of motherhood, and the risks of relationships breaking down that are putting women off having children.

> The Family Policy Studies Centre has produced a snapshot of social life in the new millennium, and you can sum it up in three simple ways: marriages down, childbirth down, solo living up. (Melanie McDonagh, 'Why Do So Many Modern Women Not Have Children?' *Daily Mail*, 28 March 2000, p. 12)

There is growing concern about the lack of support our culture provides for mothers and families. About how this is making women, especially educated, professional women, more and more reluctant to have children. Questions are being asked. 'Where will this all end?' 'What does the future hold?'

We could dwell on this issue much longer, but right now the fact remains that modern mothers are weighing up the odds and, if they have babies at all, deciding to have them later.

For some women, however, the big issue may not be *when* and *if*, but *can* I have a baby? As the next chapter will explore, for a large number of us conceiving and having a baby is not the problem; finding Mr Right is.

# 8

# What If Mr Right Doesn't Appear?

*Not finding the right person is a common cause of delayed parenthood. Some women go through the agonising process of wondering whether to get pregnant anyway and become a single mother.*

(Maggie Jones, *Motherhood After 35* (Fisher, Arizona, 1998), p. 7)

A huge concern, for many of the single women suffering from biological-clock anxiety that I met during the course of writing this book, was fear of not meeting the right partner in time. The biological clock is not just ticking – it is booming, but there is no guarantee that Mr Right will appear before midnight.

'I have had a few serious relationships,' says Jennifer, 'but nobody I wanted to start a family with. I started thinking about children when I met Roger. The trouble was he suffered from bouts of depression and didn't want children. We broke up after six years and I haven't met anyone since. I'm thirty-six now and all the decent men my age are either married, gay or chasing younger women.'

When a woman gets to her mid thirties and early forties and starts thinking about babies, life can get more complicated. Clearly the right father is a vital ingredient. 'There are no men!' single women in their thirties often proclaim to one another. This doesn't mean that there are no men, just no men these women want to have a child with. It means that women have more confidence than in the past. Our expectations for intimate relationships are rising. We don't want to have a child with any man that comes along. We'd rather be alone than in a mediocre relationship.

And it can be no easier for some women in relationships. They may be with a partner who doesn't feel ready or want children, or who has children from a previous relationship.

> 'Every time I mention children,' says Amanda, age thirty-five, 'he just shakes his head and says we should wait a few years. I love him very much so I just keep hoping he will come around to the idea. I don't want to reach that point when I have to give him an ultimatum.'

Arguing about whether or when to have a child is a major source of strife for lots of couples, not least because there are so many considerations. You may meet a partner who doesn't want children, or has children already. Or you may feel pressured by your partner to have children before you feel ready. And even should you decide to try to conceive you may encounter fertility problems.

In the past couples got married and had children, because there was simply no other option. But today it's much more complicated because of all the choices available. Should you take time off to look after the child? Can you afford a career break? How will a child affect your sex life, social life, looks, health and bank balance? With all the advances in fertility treatment, isn't there time to worry about all this in the future?

When one partner wants a baby and another doesn't, this can cause incredible tension in a relationship. There is the danger that love and commitment as a couple without children can transform into hate and resentment.

> 'I hate him for what he took away from me,' says Jill. 'For ten years he dithered about children. When I got to thirty-seven he was still undecided. Our relationship crumbled. He had an affair with one of his clients and got her pregnant. He says he wants to stay with me but still be involved with his child. Where does that leave me? I'm thirty-nine and running out of options.'

Some women feel uncomfortable deceiving their ambivalent partners, but others decide to trick their partners into getting them pregnant. Sometimes the partners do come round to the idea. But more often than not the woman is left to bring up the child alone.

> 'It was never really serious between me and Steve,' says Sarah, age thirty-nine. 'We hadn't discussed children but I was deliberately careless with contraception. When he found out, he said he was shocked but that he would stand by me. For the first few months of the pregnancy he really tried to be supportive but then, two weeks before I was due, he sent me a letter saying he really couldn't handle fatherhood. I have not seen him since.'

When a woman is up against the clock and longing to have a child this can often interfere with establishing a good relationship. The biological clock is the reason why many thirtysomething relationships become nail-biting cliffhangers and couples wonder if they should stay as they are.

I spoke to many women who felt that because of the pressure of their biological deadline they had begun to evaluate men only in terms of father potential. If the men didn't score very high they were dropped, or the need to have a child put too much tension into the relationship. The desire to have a child blinds a woman to the most obvious question she should ask herself: 'Is this really the person I want to spend the rest of my life with?'

If Mr Right doesn't appear and you are approaching forty, you may face one of the greatest dilemmas of your life. Should you give up your dream of a loving partnership to raise a child and become a single mother by choice? Or should you keep hoping Mr Right will appear and risk having neither relationship nor child?

Much depends on how essential an ingredient the father figure is.

> Amy, age thirty-six, says she wouldn't dream of bringing a child into the world by herself. 'It's just not fair on the child. It wouldn't be fair on me either. Bringing up a child is so terribly hard to do alone. My sister is a single parent and I see how exhausted, disillusioned and unhappy she is.'

> The 52-year-old actress and singer Elaine Paige feels that she could have had children but 'I'm not the kind of person who would want to do that on my own and, since I was never in a situation with a man that was conducive to having children, I have never gone there' (quoted from *Daily Mail*, 22 April 2000, p. 41).

A woman can choose to embrace singledom. In her book *Single and Loving It* (Thorsons, London, 2000), author Wendy Bristow suggests that, rather than putting life on hold, single, child-free women should live their lives to the full. But for some women, the longing for a child is so intense that they can't get on with the rest of their lives. When the 'now or never' time approaches, the decision to have a child is made, regardless of whether there is a suitable partner or not.

> Claire realised in retrospect that the man she had been married to for the last eight years fitted the pattern of all the other men in her life – he was creative and stimulating but totally uninterested in having a family. When her marriage ended she was thirty-seven, and she found herself in the same dilemma as many other women who want to have children but are alone and up against the clock. Afraid that her time to have a baby was quickly running out she began to consider an option

> which was previously unthinkable – becoming a single mother. Despite tremendous uncertainty, insecurity and little positive support from family and friends she got pregnant by donor insemination.

It is important to focus on finding the right person, giving yourself the best chance of happiness, but the biological clock can blind you to all that. You may choose to take destiny into your own hands and have a baby alone.

According to a recent survey millions of American women are setting themselves a deadline – if they can't find the 'right' man by their late thirties they aim to have a child alone. Single motherhood is becoming an acceptable option to US women disappointed at the calibre of available men.

## Single mothers by choice

In recent years there has been a remarkable increase in the number of mature, single women having babies. A powerful driving force behind this trend has to be the women's movement.

By allowing women more choice with their lives the women's movement has given women a better sense of self-esteem than ever before. We know today that there are many other ways to find fulfilment and happiness apart from marriage. We don't need the support and protection of men any more. Today marriage is not an agreement, with the wife submitting to the husband's authority, but a partnership between two equals. Men don't make all the decisions any more. Men don't bring in all the income any more. Women don't have to marry for financial security any more, but for a loving, committed relationship. A woman who doesn't marry is no longer an 'old maid'. She can have a successful and rewarding career, social and sex life and she can have a family by adopting or conceiving a child.

Women today don't have to wait around for Mr Right to appear if they want to have children. The majority of us still prefer to raise a child in a couple, but there is a steadily increasing number of women who choose to have a baby solo.

Who are these women? What motivates them to become single mothers and what gives them the strength to take this difficult and still unconventional step? 'A single mother by choice', writes Jane Mattes, author of *Single Mothers by Choice* (Times Books, New York, 1994, p. 4), 'is a woman who starts out raising her child without a partner. She may or may not marry later on, but at the onset she is parenting alone by choice.' Mattes goes on to say that this definition excludes unmarried couples, heterosexual or homosexual, who are co-parenting, and women who become single parents through divorce or death of a spouse.

Studies of single mothers by choice indicate that many are in their mid thirties and very concerned about their ticking biological clocks when they decide to become mothers. There is a tendency for them to be white, heterosexual, well-educated, career-orientated and financially secure, but generalisations can't really be made. They represent a cross section of women from all backgrounds and social and political beliefs. The only thing they have in common is a longing to mother and a firm belief that one loving parent can do a good job raising a child.

Among the women I spoke to, a very strong desire to have children was the most popular reason given for their choice. For a variety of reasons they were not able to have children within the traditional marriage context. Some wanted to get married and have children but couldn't find the right person. Others were very career-focused and couldn't find a man or woman who would work around their timetable. Others were pessimistic about relationships.

Many didn't feel that they were being particularly radical. They cited the high divorce rates and stressed how much better it is to decide to become a single mother than to be like the millions of women who are unexpectedly left with a bad marriage and an unsettling divorce.

Many felt that they didn't need a man. They had the psychological, social and financial capability to take care of a child themselves. Crucial for most was the biological clock. A number of them told me that they would not have considered having a child on their own at a younger age, but as they approached what they considered to be their reproductive deadline they were forced to make a decision.

The significant lessening of the stigma that was once attached to single mothers also made the decision easier for many women. Clearly public opinion has changed since the 1950s when Ingrid Bergman was ostracised by Hollywood for giving birth to director Roberto Rossellini's child when she was not married to him.

Single mothers by choice often commented that although people tended to assume they were not single by choice, there was growing awareness that they were mature, confident women who simply wanted to become mothers before it was too late. Some felt that they were treated with a great deal of respect for having the courage and strength to take such a bold step and that rather than being attacked for their motivations there was a genuine desire to understand them.

> 'Whenever people ask me how my little boy coped with the loss of his father I tell them straight away that he never had a real father. Of course I get the odd disapproving stare, but for the most part people are not at all shocked. Most are very respectful and understanding.' (Sam, age thirty-seven)

Should a woman consider single motherhood social stigmatisation is probably no longer her biggest concern. She is far more likely to have other worries about how going it alone might be for herself and her child. The three most pressing concerns of the single mothers I spoke to were:

1 potential consequences for the child;
2 was single motherhood right for them?
3 methods of getting pregnant.

## Will there be emotional damage to my child?

The potential consequences for the child is something that concerns everyone who is a single mother by choice. Many were aware that having parents of both sexes is still considered by many authorities on childcare to be optimal for psychological development. Children raised without a man in the house might not develop secure feelings of gender identity.

There is much controversy on this subject among experts. Some argue that children need both father and mother to mature into a well-balanced adult, others argue that they do not. That boys just 'know' they are boys and girls just 'know' they are girls, and children of single mothers can confirm their gender identity through contact with men they meet in the course of their lives. Others argue that crucial is the mother's attitude towards men. For example, negative feelings can have a negative impact on a boy's sense of maleness.

It is true that children brought up by single mothers do not have as much immediate exposure to conventional adult sex roles as other children. But some of the women I spoke to thought of this as a positive advantage. They liked the fact that their children can grow up free of gender stereotypes.

Some mothers also feel that in certain cases having only one parent can provide a greater sense of emotional security than having two. Carol Klein, author of *The Single Parent Experience* (Avon, New York, 1973, out of print), writes: 'Research has shown that despite nuclear family mythology, having a good relationship with one parent is healthier soil for emotional growth than growing up with two discontented parents' (quoted by Marilyn Fabe and Norma Wikler, *Up Against the Clock* (Warner, New York, 1980), p. 197).

Many of the single mothers I spoke to strongly agreed. Several had grown up in families where there was constant bickering and stress. Counsellors and psychologists who concern themselves with the effects on children of family stresses such as divorce and abandonment warn of behavioural problems in learning and problems establishing relationships. It is important to point out that a child born to a single mother by choice has not experienced the loss of a parent or disruption of the family.

Studies of families reveal that the crucial matter for a child's development is that they feel loved and secure and that there should be minimal family disruption – not whether or not there are two parents. Most experts in child development now think that the children of single mothers by choice have as good a chance of turning out well as any other child.

Raising a child in a two-parent heterosexual family is most people's ideal, but it is an ideal that has been over-idealised. We know that we live in a fast-changing world. The so-called non-traditional family, be it step-parent, homosexual, cohabiting, heterosexual, or single-parent family, is now more common than the traditional, and healthy, happy children are being raised in non-traditional ways.

Some child-development specialists believe that single motherhood can create an unhealthy dependency between mother and child. They argue that the child becomes everything to the mother and this is too much of an emotional burden for the child to bear. But none of the single mothers I spoke to felt that they placed too heavy an emotional burden on their children. Perhaps this was because many of them had busy active lives, stimulating careers and outside interests. They did not see the child as filling a void in their lives, only enhancing their lives. Some felt that housewives were more at risk of making their children overly dependent because of the lack of interests outside the home.

Few of the single mothers by choice were concerned about the child being socially stigmatised. They pointed out that their decision was not an irresponsible one or an accident: it had been a choice. They pointed out that in today's society single mothers were commonplace, that it just wasn't unusual any more.

Every single mother will eventually have to deal with the issue of when to tell a child about its father. The women I spoke to all handled this in their own way, depending on their relationship with the father. Some women keep in touch with the father and allow some kind of contact, and when they feel the child is ready, tell them who their father is. But what about the ones who didn't know who the father was, who had a child by donor insemination or had no contact with the father? In some cases the mother created a mythology about the father, but others saw no reason not to tell the truth.

> 'I'm proud of my decision. I've never regretted it for a minute. When Thomas is old enough I am going to tell him that because mummy wanted him so much she decided to have him by donor insemination.' (Linda, age forty-one)

The child of a single mother by choice will inevitably go through periods of confusion about his or her father. The way a woman deals with this has a great deal to say about how she feels about her decision to mother alone. Resolving her own feelings about the issue before she speaks to the child is important. Hopefully a woman can reach a stage when she has prepared herself enough and can answer questions about the lack of a father in a positive way that will make the child feel accepted and comfortable with his family situation. The 'daddy issue' is a difficult one, but it is possible to deal with it successfully.

> A child can grow up and do fine without having a dad. It may at times be confusing, but it doesn't have to be devastating, particularly if you can be helpful to your child in dealing with the issue. Empathetic listening and acceptance of feeling is invaluable. And expect the feelings to change at different stages and even from week to week. (Jane Mattes, *Single Mothers by Choice* (Times Books, New York, 1994), p. 159)

## Is single motherhood the right choice for me?

A woman considering single motherhood needs to bear in mind that the choice she makes will affect the rest of her life in the most dramatic way. It is a huge decision to make and one that really needs to be considered carefully from every angle. Advice, counselling and support during the decision-making process are advised.

If you are considering single motherhood here are some questions you might like to think about:

- Is your desire to mother an occasional yearning or something more deep rooted? Feeling lonely when you see babies may make you wish you were a mother, but think carefully about your current lifestyle and how much you value your independence and freedom.
- Can you picture yourself as a mother? Do you see yourself cuddling your baby to sleep? What is your baby like? Can you think what your baby will look like when he or she is older, or a teenager even? Do you really appreciate that the child you will have will probably be nothing like you imagined? Will you still be able to love and cherish it?
- What kind of personality do you have? Are you quite flexible and able to adapt to the demands of another person? Are you comfortable with being unconventional? Will you be able to convey to your child the concept of being different in a positive way? Are you able to cope with criticism and disapproval from others without it affecting your self-esteem?
- Can you afford to raise a child alone? Have you thought about the cost of food, clothing, day-care, education, entertainment, emergencies, illness? Money can never buy love, but you cannot bring a child into the world unless you know that you have enough money to support both of you. Do you have savings of any kind? Have you set aside money for life insurance in the event of your death or total disability, or started saving for a college fund?
- Have you elderly parents to take care of too? Will you be able to manage both them and your baby?
- Is your job very demanding? Will you be able to adjust to a more flexible schedule to suit your child's timetable?
- Have you given up your dream of parenting a child together with a loving partner from the beginning? Perhaps in the future you may meet someone, but are you prepared to accept that for now, and maybe for ever, you are in this alone?
- Have you thought about how you can meet your child's needs? About the needs of a child that only a mate can fulfil? Are you prepared for all the 'daddy questions' which will inevitably arise one day?

- Have you got a good support system? Without a partner you are going to need all the help you can get. Are you comfortable reaching out and asking for help? Many single mothers told me that they could not have survived without the help and support of friends or family or other single mothers.
- Have you thought about how having a child will affect your future relationships? What about dating? What if, in a few years' time, you meet someone you would like to commit to?

Some women, like Ellen, age forty, after much soul-searching, decided against single motherhood.

> 'I really did think I wanted a child, but bringing up a child is a tremendous job and a single mother gets little relief. If special problems come up a woman alone has no one to talk to or share the burden. I don't honestly think I could have handled it.'

On the other hand, Janice, who is raising a child alone, understood that although there were disadvantages to being a single mother, her deep desire for a child would help her cope.

> 'At least I knew what I was getting into from the start. Many women whose partners leave them don't. There are times when it is so tough, but there are wonderful times too. When I made the decision to keep my baby it was because I felt confident I could take care of her. That she would be loved. That I would do the very best for her.'

Single motherhood is a tremendous challenge. It won't work for every woman, but it does work for some. If you are approaching forty and still haven't made a decision, the important thing is to make one before your biological clock does stop ticking. Expect your feelings to change and alter during the decision-making process. This indicates that you really are examining your feelings so that, when a decision is finally reached, you can rest assured that the choice you made is right for you.

Making a decision is important because single motherhood usually works better when it is planned and prepared for than if it is an accident or you enter into it with mixed feelings. You may consider the issue and decide it is not a life choice for you. Or you may decide in your early thirties to set yourself an age limit to have a child if you haven't met the right partner. Or you may decide that the time is right for you now to go ahead with it. Whatever decision you make, at least you can be clear in your mind what you intend to do.

## How do I get pregnant?

Should you finally decide that you have the maturity, courage, resilience and flexibility to become a single mother by choice, what are your options? How do you go about getting pregnant?

There are three ways to become a single mother by choice: having sex with a man, insemination from donor sperm, and by adoption. Every woman has to decide which method is right for her. None of the solutions will be perfect and each will carry with it problems and complications. The success of each depends on how well you think you can cope with things being different from the traditional.

The obvious way to get pregnant is to have sex with the intention of getting pregnant. If you are thinking of doing this, please remember to take precautions. In this age of sexually transmitted disease it is very dangerous to have unprotected sex. Ideally you should make sure the father takes a blood test for HIV.

If you get pregnant this way the big issue is whether or not to tell the father. You may feel that it is important for the father to know, but you may also feel that you don't feel comfortable involving the father in the child's life. The relationship may have been casual, or a one-night stand, or the man could be unreliable or abusive. It is important to realise that the decision about whether or not to tell, and how to tell, the father about the pregnancy will have lifelong consequences for both you and your child. If you don't tell him there will come a day when your child will demand the right to know. If you do tell him you have no control over how much involvement the father chooses to have. If he can prove the child is his, he may get legal visitation rights.

Some women choose to become single mothers when they get pregnant by accident. If this happens, pay attention to your instinctive reaction when you heard the news. Were you pleased or horrified? Is this what you want? Coming to terms with the relationship with the father is crucial. What are his feelings about the baby-to-be? If he doesn't want to be involved, what is more important to you now, continuing the relationship or having a baby?

Much depends on how deeply involved you are with the father. How much does he want to take responsibility for the child? You may find that the news pulls you together, but you may also find that it tears you apart. You really have to think about how you feel about raising and bearing the child of this man. You have to accept the fact that this child is forever going to link you to this man. Did you have loving feelings for this man? If the relationship ended badly will you be able to cope with your feeling of bitterness and disappointment? Remember that the day will come when your child will be curious about his or her father. And you will need to present a realistic, positive picture of him, even if he was someone you had a casual relationship with. Sometimes not being very emotionally involved with the father can be a little easier. How much difficulty you experience will depend upon how ready you feel to be a single mother and what kind of involvement, if any, the father wants to have with the child.

In the UK unmarried fathers have no legal rights over their child, no say in how the child is brought up, how they are educated, what religion they are and so on. If you died custody does not automatically go to the dad. As a partner, the father is legally obliged to support your child, but he is under no such obligation to pay maintenance if the relationship flounders. According to recent research over 70 per cent of cohabiting mothers considered security for their children to be the main reason for marriage. For information and advice about the legal rights of parents contact the Family Law Consortium (2 Henrietta Street, London, WC2E 8PS).

For a small but growing number of women a way to avoid all these kinds of complications is to get pregnant by donor insemination. Donor insemination is becoming more readily available for single women in recent years. It is also used for couples experiencing fertility problems. A woman can get inseminated by an anonymous donor's sperm at a fertility clinic or she can do it herself at home.

Insemination is the method of conception when a man's sperm is introduced into a woman's vagina via a syringe. Doctors prefer to use frozen sperm to allow time for an HIV blood test. Insemination is usually done two or three times at the most fertile point of a woman's cycle when she is ovulating. Simple ovulation kits sold in chemists can tell a woman when that point in her cycle is going to occur. Insemination can be done at home or by a doctor. Some doctors place the sperm higher in the vagina than others. A woman is usually advised to rest afterwards.

Rates of success with insemination are lower than with sexual intercourse because of the hormonal changes that accompany intercourse. Donor sperm will not be refused to older women but success rates are lower because fertility declines with age. At present women under thirty achieve a live birth rate of 10–12 per cent per treatment cycle, but this begins to decrease after that age. Women aged thirty-five to thirty-nine have a 9 per cent chance of a live birth and women over forty have only a 3–4 per cent chance of a successful pregnancy for each donor cycle. Doctors will run standard fertility tests for women over the age of thirty-five who want to get pregnant.

In the USA sperm banks are far more accessible than in the UK. A woman considering donor insemination is strongly advised to get a list of licensed clinics from her GP or the HFEA in the UK or from SART (Society for Assisted Reproductive Technology) in the US (see resources section) and not from the internet or some unlicensed source. It is important that the sperm she uses has been tested for sexually transmitted disease and for suitability.

Most fertility clinics treat heterosexual couples, but some will treat single women and lesbian couples. Usually a donor will remain anonymous, but occasionally a clinic will treat a patient with sperm from a donor who is known. It is possible to request certain characteristics such as national origin and religion, but strict confidentiality on both sides is the standard. Doctors will often try to match physical characteristics as best they can so that the child resembles the mother. The cost for each insemination is several hundred pounds. Sperm donors are usually recruited from colleges or local newspapers and paid per ejaculation. Donors are asked to refrain from intercourse for two days before ejaculation so that the quality and quantity of their sperm is at its best.

One of the greatest benefits of sperm donation from an unknown donor is the simplicity. There is no worry about the father trying to locate the child and the donor does not have any responsibility for the child. If you conceive with donor insemination you need to be aware of the emotional implications for the child. You will have to be prepared for the 'daddy questions' and for your own lack of information about the biological father. It is important that you work out your feelings about the donor and about conceiving in such a non-traditional way.

If it is important to you to know the biological father of your child and to be able to tell your child about him, then you may prefer to conceive with the sperm of someone you know. If the man is agreeable to becoming a donor make sure he is tested for HIV. You may feel more comfortable conceiving the natural way, but many single mothers by choice prefer insemination because there is less possibility of emotional involvement with the father. If you do decide to take this course of action, make sure you develop a good working relationship with the man. Remember if you conceive with someone you know there are always potential legal and emotional complications throughout your child's life.

If donor insemination is not for you, you may want to consider adoption. It is still difficult for a single woman to adopt, especially in the UK, but it is getting easier.

Some women prefer to adopt rather than to conceive because adoption does have its advantages. You do have some choice over the age and sex of your child. You won't go through a pregnancy alone. If the child is older you will get lots of information about the child's background and character. Bear in mind though that your age and single status will play an important part in the age of the child you adopt. Babies are very hard to adopt. The waiting list is long and preference is always given to younger couples.

A woman who adopts a child faces different issues from a woman who chooses to have her own child. She has less control for a start over many aspects of the child's life. You won't know how well the mother took care of herself during her pregnancy. There may have been physical and emotional abuse if the child is older, and there could be severe behavioural difficulties. But then, if you had your own biological child, it comes with no guarantees. The important thing is knowing that you can love and accept your child with or without physical, emotional and intellectual difficulties. Think about how much you are going to tell your child about his or her biological parents. How you are going to cope with his or her need to know.

If you are thinking about adopting you need to be sure that you can love an adopted child as strongly as you would love a child of your own. Remember you will be taking on a huge responsibility, and do much reading and research about adoption before you make your decision. (For information and advice about adoption and fostering see the resources section at the end of the book.)

## Being a single mother

Should you get pregnant and decide to have the baby by choice, expect your world to be turned upside down in much the same way as any woman who gets pregnant. But as a single mother you will have to cope with additional stresses and strains.

It won't be easy telling the people around you. Congratulations and shock will be mixed in with disappointment and disapproving silence. The woman who is a single mother by choice is doing the unconventional thing. She doesn't fit into a neat category and people may not know how to react. Childbirth classes, hospital forms, birth certificates and so on all present areas of difficulty. But the stresses of being pregnant alone will be nothing compared to the demands of a newborn baby and caring for it all alone. Here's what Janice, single mother of Ryan, age five, says.

> 'Having my baby was the most thrilling, wonderful experience of my life. The special closeness, the wonderful memories of those early months will be forever with me. But I also remember how out of control and chaotic life was. Ryan didn't seem to sleep at all and needed constant supervision. The feeding was constant. Some nights I just cried with fatigue. There were so many times I wished I had someone else to just give me a few minutes' peace. Everything became complicated. Even going to the shops was a major expedition. I couldn't plan anything, because I never knew whether he might feed or sleep.
>
> 'I took six months off work to have my baby. It seemed like such a long time but it flashed by. I found it intensely difficult and painful to leave him with the child-minder. Our first separation was unbearable. I cried my heart out and there was no one I could share my anxiety with. Going back to work was a very difficult adjustment. Suddenly I had to stop being in mummy mode and be this calm, confident person. I felt like two different people. It is hard juggling my career with my baby. The child-minder I have is great, but she can't always be there when I need her. I can't always make my work timetable fit around my child and of course work suffers. I've definitely made less progress in my career since becoming a mother. I doubt I will ever be able to devote enough hours in the day to get on to the board of directors.
>
> 'Ryan has started nursery school now. It's tough when I pick him up alone and he sees all the other children with their dads. It's tough when he asks me who his Daddy is and I know that soon I'm going to have to tell him I don't really know who his Daddy is. It's tough when I see other dads playing football with their kids. It's tough not being able to go out at nights because there is no one to look after Ryan. It's tough doing all the worrying alone. It's tough always having to be the strong

one. It's tough not being able to share Ryan's first smile, first tooth, first word with anyone.

'Yes, it is tough, and sometimes lonely, bringing up a child alone. But it was a choice I made. And it's a choice I do not regret. I knew what I was getting into. Life isn't always easy or perfect. So far we are doing fine. He is a happy child and coping brilliantly. I'm going to make sure I keep it that way.'

Single motherhood is not an easy path to take, and many argue that a child really needs a father, but if a woman passionately wants a child and fears that she may never meet Mr Right in time, it is a course of action available to her.

Unfortunately, though, deciding to become a single mother by choice when the biological clock starts ticking does not necessarily mean a woman will be able to have a baby. Single mother by choice or not, once the decision to become pregnant has been made, the father dilemma may no longer be the problem. Not being able to conceive could be.

# 9

# Why Can't I Get Pregnant?

*As physicians have only recently realized – too late for those baby boomers who put maternity on the back burner for years – a woman's fertility may start to decline earlier than they'd thought. By some estimates, as many as a third of women who try to get pregnant after thirty-five may not be able to, among would be mothers past age forty, infertility rates rise to half.*

(Dianne Hales, *Just Like a Woman* (Virago, London, 1999), p. 197)

Infertility and biological-clock anxiety are often linked. A woman who needs fertility treatment may panic about how long she has left to get pregnant, and a woman who is prompted to make a decision by her biological clock may suddenly find herself confronted with fertility problems.

## Waiting for Baby

After the baby boom, we now have the baby dearth. Fewer babies were born in Europe last year than at any time since the

> second world war, and one in six couples now experiences fertility problems. (Tom Robbins, 'Time Bomb Alert as Births Tumble', *Sunday Times*, 16 January 2000, p. 4)

There are rising concerns of a fertility crisis. In the UK the annual increase in population has virtually ground to a halt. The number of children born each year has gone down by more than 10 per cent since 1990. According to recent figures British women are having an average 1.73 children, far below the 2.1 needed to sustain the population. The number of couples seeking fertility treatment for problems with conception has increased by 55 per cent in the past five years and now stands at around thirty thousand a year. That's around one in six couples.

Pollution, infection and the stresses of modern living are thought to contribute to the increase in fertility problems, but, as mentioned previously, modern women's choices are also driving down birth rates. More women are choosing not to have babies, and even more are choosing to delay having children until later in life, when they are more likely to experience problems with fertility. The average age of first-time mothers is now 28.9.

Waiting longer for a baby means that a woman has fewer years left to conceive. As pointed out earlier, in many cases these years are less fertile. Unlike sperm, which are constantly replenished, a woman's eggs are with her from birth. The common assumption, based on the success of donor eggs used in older women, is that these eggs lose their vitality with time and are responsible for her decreasing fertility. Another reproductive change that adversely affects a woman's fertility is that oestrogen, needed to promote menstruation and maintain pregnancy, begins to decrease in the mid thirties. Older women have less fertile cervical mucus.

The longer a woman waits, the longer she has been exposed to hazards of modern living which can affect fertility. Living in our world is not easy. We are faced with an increasing load of stress, pollutants, refined and processed foods and sedentary jobs. Evidence is also emerging that stress, tiredness, work anxiety and long hours lead to loss of libido and poor fertility among working women.

All these factors, plus the cumulative effect of an ageing body, compromise health and fertility. The reproductive organs don't live in a vacuum. They operate and co-operate with other parts of the body. If the reproductive system, or another part, of the body isn't working right, fertility suffers.

When a woman stops using contraception it may come as a surprise to her that she doesn't get pregnant immediately. She may have thought getting pregnant was easy. For some women it is, but for the great majority it isn't.

It is estimated that it takes a fertile couple having regular intercourse an average of six months to a year to conceive. It is possible to be perfectly fertile and not to get pregnant for years, especially if you haven't learned how important it is to time conception. A couple who have no apparent problems with fertility still only have a 25 per cent chance of getting pregnant after one month of unprotected intercourse. After one year of trying they have 85 per cent.

Doctors recommend that a woman under thirty waits two years before she seeks fertility treatment but, in recognising the wear and tear our bodies go through with age, they are advising older women not to wait so long for a baby and to seek help much sooner. Once infertility treatment is begun there really is no telling how long it will take for a woman to conceive or even if she will be able to have a baby.

Many women, aware of their biological clocks ticking, are taking their doctor's advice and starting infertility treatments after six months of unsuccessfully trying to get pregnant. Chances are, by this time they will already be dealing with the pain of infertility.

## The pain of infertility

Almost all women assume that they will be able to have children someday, even if they are not sure they want to; the potential to have them is important even to women who never intend to have children. The ability to conceive and bear children is

> an almost instinctive birthright. (Christiane Northrup, *Women's Bodies, Women's Wisdom* (Bantam, New York, 1998), p. 350)

Coping with infertility is a life-changing event. It is especially hard if there have been months or years of infertility treatments giving a false sense of hope. About two-thirds of women who undergo fertility treatment do have a successful outcome, but a third don't.

A woman who decides not to have children knows that she could have children if she wanted. She has a choice. But a woman who has said yes to children, but her body says no, has been robbed of the privilege of choice. She often feels great bewilderment and a sense of injustice. She asks 'Why me?'

> Infertility is a kind of death. However unlike death, which whilst deeply painful to confront is part of a natural process, not being able to procreate when you want to feels cruelly unnatural. It is as if Nature has cheated us out of our birthright and we don't know how to cope with the rush of emotions that plague us. (Anna Furse, *The Infertility Companion* (Thorsons, London, 1997), p. 119)

There may be feelings of abnormality, difference and, worst of all, failure. The body has failed to deliver. Everywhere you go the world seems to be full of babies and mothers. Most women who undergo years of treatment have thought long and hard about the decision to have children. They know what it means to mother and not to mother, and for most of them not mothering is a living hell. They may wonder what the point of their lives has been. They may not feel worthy, creative and whole as women. They may feel guilt about the decisions they made in the past, especially if they postponed for too long. They know that they have to restructure their life and move on, but sometimes the pain is so intense they cannot see a way forward.

> 'I think, what is the point of my life? I feel I've let down everyone.' (Sarah, age thirty-seven)

> 'Everywhere you go there are reminders of what you couldn't do, but the rest of the world seems to be able to quite naturally. I see babies and it tortures me.' (Lucy, age forty)

> 'I suppose this is a wonderful opportunity to develop myself in other areas of my life. But the thing is I don't want to. I'll never have my own baby and nothing is going to take away that pain.' (Rebecca, age thirty-nine)

The welter of depressing emotions that arise when infertility is diagnosed is not a sign of weakness. It is a well-known fact that the stress of infertility can be devastating. This is because infertility triggers so many emotions, hurts our self-esteem and alienates us from other people's lives. Infertility treatment can make or break couples and is a real test of the strength of the relationship.

Infertility treatments bring a heightened state of awareness to a process the majority of us take for granted. It seems our God-given right to the miracle of reproduction allows some people more rights than others.

Fertility is a gift we are given, it is not a right. Talking to infertile women and couples in the process of writing this book made me appreciate how important it is for a woman not to take the decision to mother lightly. With so much female identity resting on motherhood, it is hardly surprising that women who want to have children and can't, feel an unbearable sense of loss.

## Causes of infertility

Pregnancy, even for a fertile couple, is not as simple and easy to achieve as we often think. In fact, considering how complex an event conception is and how many factors have to synchronise, it is surprising that it occurs as often as it does. At any stage in the intricate process something could go wrong. The pituitary gland could malfunction and hormones may not be released or released at the wrong time. Ovulation may not occur. If an egg is released, sperm may not meet it in time. The fertilised egg may not implant in the womb or it may be rejected by the mother's body. An abnormality in the fetus or in the mother's hormone levels may result in miscarriage.

Conception really is a lottery. There are no guarantees, even for a fertile couple. And if there are problems with fertility the odds of achieving a healthy pregnancy are less in a woman's favour.

Age is the most significant factor. As discussed earlier, fertility declines with age. The most fertilisable eggs are released earlier in life. Although the causes of infertility, listed below, do occur in young women, their incidence is far higher in older women.

At least 20 per cent of women attending a fertility clinic will have problems ovulating. Without an egg released during the monthly cycle pregnancy cannot occur. Ovulation is triggered by the release of hormones from the pituitary gland in the brain. Lack of ovulation is normally caused by the failure of the body to produce enough pituitary hormones or they are released at the wrong time. The body is placed in a state of hormonal imbalance.

Pituitary malfunction is often the result of problems with the hypothalamus gland, also located in the brain. The pituitary is controlled by the hypothalamus, so anything that affects the hypothalamus can affect this gland. The hypothalamus is sensitive to severe physical and emotional stress. Every woman who has missed a period due to the stress of work, intense exercise, poor diet, weight loss or gain, travel or illness, will know this.

Once a woman passes the age of thirty regular ovulation every month is less likely to occur. At menopause, which usually occurs when a woman is in her late forties or early fifties, ovulation stops altogether when the ovaries stop working or run out of eggs. Premature menopause is rare, but it can happen to women at any age. It can be caused by infection or illness, but it may also be that some women are born with fewer eggs than normal or have a tendency to discard more eggs each month.

Polycystic ovary syndrome, where the cysts result from hormone imbalance, is another very common reason for ovarian failure. The condition tends to affect women in their thirties. High levels of the hormone prolactin, which is normally produced during breastfeeding, is a rarer condition, but it can inhibit ovulation too.

Approximately a third of female infertility problems are due to damaged fallopian tubes from some infection, illness or surgery or abnormalities present from birth. Untreated sexually transmitted diseases, like gonorrhoea, can result in infertility. Unfortunately many women don't have severe symptoms and may not realise the damage that is being done to their fallopian tubes. Pelvic inflammatory disease, or PID, can start after an induced abortion, miscarriage, childbirth or surgery in the pelvic region or after an infection with a sexually transmitted disease. Other infections which can affect fertility include the silent epidemic, chlamydia. This bacterium, which resembles a virus, often doesn't have any symptoms and it can destroy a woman's fallopian tubes. It is the fastest growing sexual infection among young people, with 38,000 cases reported in 1997. Previous surgery in the abdominal region can damage the tubes too. Bleeding or injury to the tissues may also cause adhesions to form.

Endometriosis is thought to cause about a fifth of female infertility. This is when the womb lining starts to grow in other parts of the abdomen where it creates scarring which distorts the reproductive organs and stops them working properly. The condition is most likely in women in their thirties and forties who have not had children.

Uterine disorders and fibroid tumours affect fertility. About one third of all women have fibroids or polyps over the age of thirty. They are benign swellings which don't often cause symptoms, but women with fibroids may find that they can't get pregnant. Abnormalities in the cervix and the womb can also prevent a pregnancy occurring. The womb may not be able to support the pregnancy and the cervix may not be able to hold the baby in the womb.

In rare circumstances an embryo will start to develop outside the womb, usually in one of the tubes. It can happen if a woman became pregnant with a coil in place. It can also be caused by a damaged tube which stops the embryo moving into the womb. An ectopic pregnancy, as it is called, is very dangerous and causes severe pain. It must be surgically removed as quickly as possible before the tube bursts and too much damage is done.

A woman may also find that she has problems conceiving when she stops taking contraceptives. Use of condoms, diaphragm and cap won't affect fertility at all, but the intrauterine device (IUD) can increase the risk of infection and the contraceptive pill sometimes leads to amenorrhoea – in which a woman's periods don't return when she stops taking the pill. This post-pill amenorrhoea can last up to two years. Contraceptives method are only rarely a cause of infertility. But as a woman gets older, using contraception could mean that she is less fertile by the time she stops. Using the pill can also disguise fertility problems for years.

One in five pregnancies ends in miscarriage, most of them happening in the first twelve weeks. The risk of miscarriage again increases with age. Doctors don't really know why miscarriages happen, but suspect some abnormality in the egg or sperm, hormone deficiencies, immune problems, illness and stress, environmental toxins and abnormalities in the womb or cervix. Repeated miscarriages for the older woman who feels she is running out of time can be devastating. Many feel incredible guilt and blame themselves, but more often than not miscarriage is nature's way of getting rid of conceptions that would not result in a healthy baby.

About a third of infertility problems are female and a third are male. Another third are caused by a combination of factors in both partners, such as a woman's producing sperm antibodies to her partner's sperm. In about a fifth of these cases the cause is 'unexplained'. This could just mean there is no problem and the couple have been unlucky. It could also mean there is a problem, but nobody knows what it is.

## Psychological factors

Infertility is never completely straightforward. Many factors, physical, psychological and emotional, are involved, 'so many that it is ridiculous to try to reduce fertility to a matter of injecting the right hormone at the right time', writes Dr Christiane Northrup in *Women's Bodies, Women's Wisdom* (Bantam, New York, 1998), p. 351.

There are many women, who have been told by specialists that they have 'unexplained' infertility, who get pregnant without any treatment at all.

Conventional medicine will focus on the body as the cause of infertility and it will often ignore the part emotional, psychological and even nutritional factors play in conception. An increasing number of doctors, like Northrup, are placing more importance on how a woman's state of mind can affect her fertility. Others believe that sexual intimacy and orgasm can increase a woman's chances of conception. Fertility treatments won't often take this into account.

Many infertile women are highly motivated career women trying to squeeze fertility treatment into their busy schedule, but conception is a delicate, unpredictable process, and it can't be scheduled. There is a well-known connection between high levels of stress and lack of ovulation. 'Conception leave' when a woman takes time off to prepare her body for pregnancy is an option some working women, unable to conceive, are starting to take.

Some experts believe negative beliefs affect fertility, such as, negative feelings about being female, aversion to sex or an unhappy relationship. Several studies of women who repeatedly miscarry indicate that there may be an interplay between emotions and the hormonal systems involved in pregnancy. Sometimes there are problems accepting motherhood, or the feminine role, or there is repressed anger about other people's demands, or severe emotional conflicts about motherhood.

Such theories have never been proven, however. It is just as useful to say that fertility, like all other body processes, goes better when we are relaxed and happy. If your cause of infertility is 'unexplained', perhaps decreasing stress will help you conceive. It is always in your best interests to live at a pace of life and a style of life that brings the most satisfaction.

## Helping yourself

Many women with fertility problems are not completely infertile. There is usually some chance, even if it's small, of conceiving naturally. There are also ways a woman can help herself.

Avoiding stress and leading a healthy lifestyle will optimise chances of fertility (see Chapter 11). Assessing how much you really want a child can avoid emotional stress which can inhibit fertility (see Chapter 5). Most important of all, perhaps, is being aware of the importance of timing when making love.

A normally fertile couple is able to have sex as and when they feel like it and have a reasonable chance of conceiving. A subfertile couple needs to be a bit more calculating.

A woman is only fertile when she ovulates. This usually happens fourteen days before the start of her next period. Unfortunately few periods arrive on the dot, so it can be hard to know when ovulation occurs. Although sperm can survive up to five days in a woman's body, an egg only stays viable for twelve to twenty-four hours after ovulation. So how can you tell when is the best time to have sex?

In the past this was a fiddly business involving the woman taking her temperature each morning, but on the market now are 'ovulation prediction tests'. They can tell you whether you are likely to ovulate in twenty-four to thirty-six hours. If you have sex any time during the next two to three days you maximise your chances of getting pregnant. Another device is the 'baby comp', which is a combined thermometer and mini-computer. It can tell you when you are likely to be most fertile. A simpler, but less precise way, to determine the fertile period is through vaginal mucus. This may start to feel damp a few days before ovulation. At first it will be white or yellow, then cloudy and finally clear. The clear, wet mucus coincides with peak fertility. Some women notice other bodily changes during this time which include lower abdominal pain, increased libido, a slight discharge of blood or spotting, and breast tenderness.

Vaginal mucus indicates not only ovulation, but fertility in general. When mucus is absent, pasty, crumbly or has a dry infertile quality, then sperm are unable to survive long enough to fertilise the egg.

The naturally acidic nature of the vagina kills sperm and protects the vagina from infections. When cervical mucus is wet, slippery, clear and stretchy as it is around ovulation, it will nourish and protect the sperm for up to five days.

How the cervix (opening of the womb) sits inside the vagina also plays a role in fertility. During infertile times the cervix positions itself high in the vagina, its opening is narrow and the cervix feels hard and unreceptive. As the fertile phase approaches the cervix lowers itself into the vaginal canal, softens and opens, inviting sperm to enter.

If you do not care to monitor all these signals, the simple advice is to have sex every other day, bearing in mind that the most fertile period is around the middle of the cycle – that is between nine to fifteen days after the start of the woman's last period in a 28-day cycle. Every two days is best because it takes a man about two days from ejaculation to regain sufficient sperm levels. Any longer and the proportion of older, less effective sperm is greater. (Women suffering from endometriosis might need to time sex slightly earlier because the sperm has a more difficult journey to make.)

Doctors disagree as to whether any of the following techniques make a difference, but they might be worth considering:

- Avoid dairy products. Studies have shown that in countries where milk consumption is highest, women experience the highest age-related drop off in fertility.
- Avoid aspirin which can hinder egg production.
- Take cough syrup to thin cervical mucus and make it more sperm-friendly (1–2 tsp daily from day five of your period till ovulation). Make sure the product only contains guaifenesin without added decongestants or antihistamines which can dry out cervical mucus.
- The man shouldn't have a hot bath, which will suppress sperm before sex.
- Lubricants can kill sperm.
- The best position for bringing the sperm in contact with the uterus

is with the woman on her back with a cushion under her spine.

- If the woman has an orgasm this will help draw sperm up into the cervix.
- The penis should stay in the vagina till it is flaccid so as not to pull sperm out too soon.
- The woman should not douche the vagina or urinate immediately after sex.
- The woman should stay lying down for about half an hour afterwards.
- Living in artificial light without going outside in the natural light can also adversely affect fertility because light is a nutrient.
- There is also evidence to suggest that if a woman's bedroom is penetrated by light during her nightly sleep her menstrual cycle and perhaps her fertility will be affected.
- Fertility also varies with the seasons. There is evidence that human sperm and egg are more receptive to fertilisation during particular times of the year. Sperm count and volume increases between February and May. In women the quality of her egg and uterine lining is highest from October to March, peaking in November.

Sometimes simple lifestyle changes and an awareness of timing can result in a pregnancy. But a woman over thirty-five may not feel that she can afford to experiment. She knows that she has significantly less time to get pregnant than a woman in her twenties. Rather than adopting the 'wait and see approach' she usually prefers to seek fertility treatment from a clinic.

# 10

# Fertility Treatments

*For millions of men and women, starting a family is a long and distressing process . . . These people, many of whom have postponed childbirth to further their education and careers, need expert answers concerning the diagnosis and treatment of infertility. No longer do infertility patients say to their gynecologist, 'Let's get pregnant.' Couples want to know what is wrong and how their doctors are going to help solve their problems.*

(Robert Franklin and Dorothy Brockman, *In Pursuit of Fertility* (Henry Holt, New York, 1995), p. ix)

Before registering with a fertility clinic a woman should first make sure that it is approved by the HFEA (Human Fertilization and Embryology Authority) in the UK. The HFEA is a national regulatory body established by Parliament in 1991 to oversee the licensing and good practice of all UK fertility clinics both in the NHS and the private sector. The HFEA is also a good source of information and advice. (See resources section for HFEA address and for useful addresses and contacts about infertility.)

The HFEA will provide a list of fertility clinics in the UK, and then it is a matter of contacting them to see whether you can be treated. Most take on both private and NHS patients, but bear in mind that the NHS is reluctant to treat women over the age of thirty-five. The HFEA has not set a national age limit, but some clinics have upper age limits for women they are prepared to treat. The majority still only treat heterosexual couples, but an increasing number are changing their policy and treating single women and gay couples. Subfertile single women have fewer options available to them from fertility clinics than subfertile women in couples.

In the US government agencies and professional organisations like the American Society for Reproductive Medicine (ASRM) and The National Advisory Board on Ethics in Reproduction (NABER) provide informational material for patients about procedures and clinics offering fertility resources, but there is no national regulatory body. Before a woman seeks treatment in the US it is crucial that she makes absolutely sure that she is working with qualified, experienced professionals.

When visiting a fertility clinic a woman and her partner, if she has one, will be given a routine medical examination to check that the reproductive organs are normal. The doctor needs to determine whether it is the man or the woman, or a combination of the two, that is causing the problem.

One of the first tests a woman will be given will be to find out if she is ovulating. The doctor may prescribe a three-month temperature test or order a progesterone blood test which is taken around day 24 in a 28-day cycle and is a good indicator of whether or not ovulation has occurred. Blood tests at various points in her cycle will also help a doctor determine how normal hormone levels are and if there is an imbalance. If a woman is not menstruating she will be given a dose of progesterone to induce a bleed and to see if she can ovulate.

A postcoital test may also pinpoint why a woman is not conceiving. The woman makes an appointment around the time she may be ovulating. The couple is asked to have sex the night before the appointment. At the clinic a sample of the womb's cervical mucus is analysed. The doctor can tell from the quality of mucus if a woman has ovulated. It is also possible to tell if the sperm is normal and whether or not it is surviving in the woman's mucus.

Doctors may also want to take an X-ray of the uterus and fallopian tubes. Ultrasound may be used to assess the health of the ovaries and the womb. A laparoscopy is sometimes used to detect abnormalities in the womb and ovaries. This involves a small incision in the navel under general anaesthetic which can help the doctor see if the organs are damaged in any way.

If any of these tests can determine the cause of problems with fertility, such as ovulation failure, problems with cervical mucus, or damaged fallopian tubes, then treatment will follow.

## Drug treatments

Oestrogen pills or vaginal cream can be given to women to stimulate mucus production and make the mucus more sperm-friendly. It doesn't always work though, and can disrupt sperm production. Synthetic progesterone may also be given by a vaginal pessary (tablet you insert in the vagina) or by injection, but this doesn't often produce enough progesterone after ovulation for the womb lining to build up. Steroids are sometimes given to women to suppress sperm antibodies in their cervical mucus. Steroids have serious side effects and are given only for short periods.

Far more effective are the fertility drugs for women unable to ovulate. The first and most famous of these that triggers the release of FSH (follicle-stimulating hormone) and LH (luteinising hormone) in the pituitary is clomiphene citrate, better known as Clomid. Clomid is often used for women suffering from polycystic ovary syndrome. It can also help women with progesterone deficiency. Clomid is taken in pill form for about five days of each cycle and it induces ovulation in about 80 per cent of women although only 30 to 40 per cent will get pregnant as a result. This may be because the drug also causes cervical mucus to thicken. Oestrogen is sometimes given to counteract this.

Clomid can make a woman feel sick or giddy, give her hot flushes and nausea and cause pain by overstimulating the ovaries. There is also an increased risk of twins. Recently there has been concern that it might increase the risk of eggs being released with chromosomal abnormalities. Others think there might be long-term effects on the children conceived from fertility drugs, but there is no evidence whatsoever to suggest this. Clomid has been in use for years and is considered safe, if used in moderation.

If a woman has produced an egg, but can't release it on Clomid, HCG (human chorionic gonadotrophin) will be injected to induce ovulation around day fourteen of the cycle.

If Clomid doesn't work, Pergonal or HMG (human menopausal gonadotrophin) may be used. This is a hormone extracted from the urine of pregnant women. It stimulates the ovarian follicles containing the egg. HMG is usually given as a daily injection followed by an injection of HCG to trigger ovulation. About 90 per cent of women do ovulate as a result, but not all will get pregnant. The risk of multiple pregnancy is about one in five. HMG is a potent hormone and the serious risk is overstimulation of the ovaries which can permanently damage ovaries. To avoid this a woman must be monitored daily. A new pump-like device can be attached to a woman's arm and provide the correct dose of hormone through a needle. But this can be painful and unpleasant.

The drug most often used to treat endometriosis is called Danol (danazol) and it is taken in pill form. At least a fifth of women experience side effects which include menopausal symptoms, excess hair growth, weight gain and greasy skin. You can't get pregnant on the drug and it doesn't completely cure the condition, but some women do conceive after the treatment.

Women who can't ovulate because of high prolactin levels will be given Parlodel (bromocriptine). Parlodel prevents the pituitary from producing ovulation-inhibiting prolactin. This can make some women feel very sick but it leads to pregnancy in many cases.

Fertility drugs to induce ovulation do have a high success rate. However, sometimes they don't work and more advanced treatment, like in vitro fertilisation (IVF), will be the next stage. We don't know why some women respond better to the drugs than others. Also, only a certain number of treatment cycles on fertility drugs are advised, to protect against the risk of ovarian disease.

## Other treatments

Other treatments may be resorted to, if drug treatment isn't effective. For polycystic ovary syndrome there are two surgical treatments. The first involves cutting a wedge out of the ovaries, and the second burning off the cysts. With surgery through a laparoscope or microscope both are now fairly successful.

For blocked fallopian tubes there is a fair chance of correction depending on the type of damage. However, not all types of damage are suitable for surgery and after surgery the risk of ectopic pregnancy increases.

If a woman is producing sperm antibodies in her cervical mucus a new treatment is being tried by doctors as an alternative to steroids. This is to remove the infected mucous glands by laser and then use drugs to stimulate new ones to grow. Fibroids can be removed by surgery but the pregnancy rate after treatment over the age of thirty-five is only 35 per cent.

Surgery is also used for abnormally shaped wombs. If a woman's cervix is too weak to keep the baby in the womb this is called an incompetent cervix by doctors and it can be stitched shut early in the pregnancy. If endometriosis is severe, this can be corrected by laparoscopic laser surgery or microsurgery, but symptoms may return. Conception rates after surgery are around 70 per cent. If the endometriosis is minor, this reduces fertility by half compared with normal and treatment doesn't really help.

If a woman has run out of her own eggs, or there is consistent abnormality in the eggs she produces, she may consider using eggs from a donor. Egg donation is not a simple procedure because another woman has to be involved. The egg donor will need to undergo superovulation and uncomfortable surgical procedures.

Egg donors need to be highly motivated to agree to the procedures. Some women who have had fertility problems themselves donate eggs. Relatives and friends may also wish to do so. The recipient is prepared for pregnancy with hormones and the eggs fertilised in the laboratory and then introduced transvaginally as in IVF (see below).

The whole procedure raises ethical questions and it can stir up strong emotions in both donor and recipient. In the UK the sale of human eggs is illegal. A woman can't be paid and as a result there is a shortage of donated eggs. In the USA, however, egg donation is a lucrative business. A worrying trend is that there seems to be a preference for tall, beautiful, athletic and intelligent women. But beauty comes at a price. Shelly Smith, an LA-based egg broker pays the donors she uses between $300 and $50,000 for their services and charges infertile women even more.

## Becoming an egg donor

To find out more about egg donation, you can contact the Human Fertilization and Embryology Authority (HFEA) or CHILD or CARE. It can be comforting to know that you are helping other women who have not had the option to decide about babies.

It is important that you are utterly convinced you are doing the right thing before you contemplate egg donation. Egg donation is painful and time-consuming and may carry a danger to your health. Some clinics extract eggs under general anaesthetic, which is a risk, and studies have suggested ovary-stimulating drugs may trigger ovarian cancer. You will have to have daily injections for two weeks and there is a slight risk of infection when the eggs are extracted which could damage your fallopian tubes. Unless you donate to a friend or relative you won't know who receives your eggs. You may be told if a pregnancy results but you won't be told anything about the child.

## Male infertility

If a woman has a partner it is possible that he is the cause of the infertility. Male fertility also declines with age, but much more slowly and far later than in women. If both partners are a little older this combination of slightly lowered fertility can make pregnancy harder to achieve.

Men will be asked to present several sperm samples to check that the sperm is not causing the problem. Blood tests may also check hormone levels. The doctor may want to do a testicular biopsy. This is when a piece of tissue is taken from each testicle under general anaesthetic to check sperm production and find out if there is any tube blockage. In some cases an operation called a vasography to check there is no blockage of the vasa deferentia (the tube that transports sperm out of the testis) will be required.

The vasa deferentia can be blocked from birth because of a defect, by sexually transmitted diseases and through surgery as in vasectomy. Blocked tubes can be restored surgically, but there is only a 50 per cent success rate. A man with blocked tubes tends to produce antibodies to his own sperm because they cannot be ejaculated. There are procedures available now which can remove sperm from the testes to fertilise the egg.

Undescended testes, if not diagnosed early, can also cause male infertility, as can infections of the testicles and variocele, a 'varicose vein' of the testicle. Variocele can be treated by a simple operation to tie off the vein and it improves sperm quality in about two-thirds of cases. Ejaculation disorders can also cause complications, for example, when a man ejaculates backwards into the bladder. Having sex with a full bladder is sometimes recommended.

But by far the largest proportion of male infertility problems are caused by low sperm count or abnormalities in the sperm. Sperm counts have dropped dramatically since the 1960s owing to changes in diet, higher stress and pollution levels. Average sperm counts in the UK have fallen by almost half since 1938 and are continuing to decline as fast as 2 per cent per year.

Variocele, infection, autoimmunity, hormonal and congenital problems may be to blame, but more often than not age, health and environment create abnormal sperm. Various hormone treatments, such as Restandol which contains testosterone, have been tried but success rates don't tend to be high. Antibodies are sometimes tried for sperm problems. Steroids may be prescribed if a man produces antibodies to his own sperm. Many of the drugs have unpleasant side effects, such as loss of libido or loss of body hair.

The split-ejaculation technique may help men with low sperm count. This is when the richest part of several ejaculations are pooled in a lab and inserted into the vagina through artificial insemination. IVF treatments, which mix the sperm directly with the egg so that it doesn't have to make the arduous journey to it, can also help subfertile men.

Working in a very hot environment can cause sperm problems, as can illness, especially if the man has a fever. Doctors will usually advise men with sperm problems to avoid hot baths, wear loose-fitting underwear and trousers, eat a more healthy diet, stop smoking and drinking alcohol and to spray their testicles with cold water twice a day. Sometimes, though, this is not enough to make a difference.

If sperm count is consistently low or non-existent and conception impossible to achieve, the usual alternative is artificial insemination by a donor or donor insemination (DI). Donated sperm are usually introduced into either the woman's cervix or womb after ovulation.

## Assisted conception

The drugs, procedures and tests described can take months or years to achieve conception, if it occurs at all. There are also uncomfortable side effects. By the time a couple considers assisted conception they may have already put themselves through much. But they often find it difficult to stop. The longing for a child has become so great now that they are prepared to sacrifice anything.

In vitro fertilisation (IVF) is the original 'test tube baby' technique first done successfully in 1977. An egg or eggs are removed from the woman under local anaesthetic. They are mixed in the laboratory with fresh sperm and left for a few days. A maximum of three fertilised eggs will then be placed in the woman's womb through the vagina and cervix.

The technique was developed for women with blocked tubes, but is now used for endometriosis, cervical mucus problems, sperm disorders and unexplained infertility. The technique is expensive in both the UK and the US (cost is currently around £1,700). Some clinics have an age limit of thirty-five. The age limit is based on the fact that the chances of a successful pregnancy using IVF diminish swiftly – but not totally – as a woman approaches forty. Usually, though, if a woman is paying privately there is no age limit. IVF gives a couple a 10–15 per cent chance of a live birth, less if the woman is over forty, overweight or a smoker and her partner has a low sperm count.

For women with healthy tubes a simpler and cheaper method is called Gamete Intra Fallopian Transfer (GIFT). This is similar to IVF but the eggs are collected from the ovary with the woman under general anaesthetic and then immediately placed in the tubes together with the sperm. The embryo is formed not in a culture medium, but in the woman's own tubal fluid. This means that success rates are slightly higher than IVF.

Zygote Intra Fallopian Transfer (ZIFT) may be helpful for older women. It is a combination of IVF and GIFT when embryos are placed in the tubes. For cervical mucus problems, mild endometriosis, low sperm count and unexplained infertility, Intra Uterine Insemination (IUI) has about a 10 per cent success rate. A sample of fresh semen is placed in the woman's womb after superovulation through the vagina without anaesthetic. For severe male sperm problems Micro Assisted Fertilisation treatment (MAF) is used. A few or even just one sperm are injected into an egg. The woman goes through the usual superovulation and egg-retrieval procedure. There are concerns that MAF treatments increase the risk of a child born with abnormalities.

Older women who can't use eggs of their own may have greater chances of success using a donor egg. Although the child a woman gives birth to is not genetically hers, when healthy donor eggs are used conception is more likely to occur. Donor eggs are used for women who have premature menopause or for post-menopausal women. Provided the woman has no abnormalities in her womb the success rate is high. Success rate tends to depend more on the age of the donor than on the age of the woman giving birth. Following the huge success of egg donation in the 1980s and early 1990s many doctors are using the same treatment on women in their fifties.

## Other options

One option for a woman who can't conceive or carry a child herself because of age or an abnormality of the womb is surrogate motherhood. A surrogate mother will carry a child for another woman. She will either be a host surrogate – where the infertile woman's egg is fertilised with the partner's sperm and introduced into the surrogate mother's womb – or a straight surrogate – where the surrogate's egg is used with the partner's sperm.

As with donor insemination and egg donation, the USA is far less squeamish and much more businesslike about baby-making than the UK. Their fee-paying system, professional psychological support and legally bound commitments for all parties is proving to be very successful for infertile couples, women seeking sperm donors and surrogate mothers.

In the UK, until recently women who became surrogates could expect to receive around £10,000 in expenses. Although it is illegal for a surrogate to receive fees, she could be compensated in expenses. Recent government deadlines, however, have tightened up the whole business and a woman will receive virtually no financial reward at all. As a result, finding a suitable surrogate is very hard indeed, unless there is a willing friend or relative.

Lack of government vindication has forced the founder of the COTS (Childlessness Overcome Through Surrogacy), Kim Cotton, to shut down the charity she set up to help, advise and support infertile couples and surrogates and enhance public awareness of surrogacy. She felt that she was fighting a losing battle and now, without the backing of COTS, the future of surrogacy in the UK looks even more uncertain.

Surrogate motherhood is controversial to say the least. The famous 'Baby M' case in New Jersey, USA, hit the headlines when a surrogate changed her mind and took the baby back. In the UK Greg and Deborah White paid Diane Richardson £10,000 for a straight surrogacy, but Richardson refused to hand the baby over at birth. Surrogacy has its success but more often than not it is traumatic, uncertain and difficult for all parties involved.

Another option for those who cannot have their own babies is adoption. In the US private adoption is legal but in the UK it can only be carried out by local authority social services departments or voluntary services like Barnardo's. The process involves a medical and police check and all aspects of your life will be checked by social workers. Unfortunately older mothers tend to have difficulty adopting a baby – the upper age limit for babies is around thirty-eight – but may be able to adopt an older child. There may be a lack of babies, but there is no lack of older children in care who need families. These children may have behavioural and social problems, but many people who adopt them find great fulfilment in becoming their parents.

Some British parents, finding the rules and regulations of adoption so restrictive, have adopted from overseas where a large number of babies are in orphanages. The laws of the countries from which they adopt may be less restrictive, but this varies from country to country. The process of adopting abroad is similar, but you will pay for legal and travel expenses. Your local social services will provide information.

Fostering is very similar to adoption. You take care of a child's needs and integrate that child into your lifestyle. The difference is that when you foster a child you don't become their legal parents and usually share the care of the child with the parents. You may also only have the child for a short period of time. You will probably receive an allowance for the care of the child and the relationship you have with a child will be monitored by social workers. (See resources section for contact addresses.)

The US has an organisation called Big Brothers/Big Sisters which provides professionally supervised adults to serve as role models and mentors for children from single-parent families.

## Alternative therapies

In recent years there has been a proliferation of alternative medical practices now generally called complementary because they are used to complement or support conventional treatment. Briefly mentioned below are a few of the most common and admired treatments for infertility.

Herbal medicine, whether Chinese or Western, has been used for thousands of years. In China, it is combined with acupuncture, the official mainstream form of medicine. Many drugs come from herbs but herbs have less of a toxic effect. This is not to say herbs are totally without danger – some are very strong – and they should only be taken under expert guidance. Herbs, like ginseng for men, or squaw vine for the womb, can boost the reproductive organs and increase sex drive. Others, like wild yam, can act like hormones or balance hormone output in the body. Studies of oriental herbal treatments do show that they can be effective for conditions like endometriosis, ovulation problems, sperm abnormalities and unexplained infertility. Damiana, dong quai, false unicorn root, ginseng, gotu kola, licorice root and wild yam root are also good for women.

Acupuncture is an extremely effective way of balancing the body. It involves the insertion of small needles just below the skin's surface. Like herbal medicine, acupuncture works to stimulate the body's healing responses and can promote a general sense of physical and emotional well-being.

There are many thousands of herbal doctors and acupuncturists working in the UK and US. To get a list of qualified practitioners in the area near you, contact the British Acupuncture Council or the National Commission on the Certification of Acupuncture.

Reflexology is an ancient practice which works along the same lines as acupuncture, but has its own system of energy lines. Reflex points in the feet and hands correspond to parts of the body. To the reflexologist the soles of the feet are like a map to the rest of the body. Applying pressure and massage to the reflexology energy lines will promote healing and relieve stress. Certain reflex points correspond to the reproductive organs and hormonal system.

Homeopathy is one of the more established systems of natural health care. A growing number of doctors are trained in homeopathy and several hospitals offer it alongside conventional medicine. It is based on the principle of 'like cures like', a bit like a vaccination. The sick person is prescribed minute doses of substances that would produce in a healthy person effects similar to those occurring in the disease. This is thought to stimulate the body's own healing powers. Remedies are made from ground-down plants, minerals, metals and other substances. The substance is diluted with water and alcohol many times and with each dilution the mixture is shaken vigorously. The more the mixture is shaken the more powerful it is said to become.

There have been scientific studies of homeopathy and infertility and it has been shown to be especially helpful for hormone problems, sperm disorders, endometriosis, fibroids, infections and unexplained infertility.

Massage is one of the oldest forms of healing and a popular natural therapy. It can improve circulation, relax muscles, aid digestion and speed up the elimination of toxins. It is also a stress-buster and strongly advised for those undergoing infertility treatment. Of special benefit to the infertile woman is aromatherapy massage.

Naturopaths are the general doctors of natural medicine. Naturopathy is based on the ideas of nineteenth-century European doctors who revived ancient Greek principles of good health based on the importance of clean water and air, good food, exercise and relaxation. The therapy uses general dietary improvements, nutritional supplements and herbs to cure illness. An increasing amount of research is showing the importance of nutrition in fertility. For example men suffering from sperm problems show significant improvements if given courses of vitamin E, B12, C and zinc. Studies also show that women suffering from infertility can be deficient in iron, vitamin C, E, calcium, magnesium, selenium and zinc.

Aromatherapy is a natural treatment which uses the healing qualities of essential oils taken from flowers, leaves, stems, plants and trees. The essential oils can be absorbed into the bloodstream via the skin, hair and lungs (through inhaling). Oils in the bloodstream will circulate for several hours, while those inhaled will stimulate the limbic system.

Oils of particular relevance in helping infertility problems include clary sage, geranium, jasmin, melissa, rose and ylang ylang, all of which can regulate the hormonal system, among other benefits. To find a qualified aromatherapist contact the Register of Qualified Aromatherapists.

Reducing stress, getting enough sleep, exercising regularly so that your muscles get stretched and your blood circulates, can all prove beneficial for infertility. Forms of exercise recommended include walking, swimming, yoga and tai chi. Meditation can also promote a calmer and less anxious response mechanism to the stresses of daily life.

A sensible, balanced diet is also recommended for anyone suffering from fertility problems. Not only does this promote general health but it is particularly advised when trying to get pregnant. Chapter 11 contains information about a balanced diet and tips for healthy eating to promote fertility.

Foresight – The Association for the Promotion of Preconceptual Care – was founded in the late 1970s in the UK with the mission to pioneer preconceptual care through nutritional medicine. Vigilant in its resource and thorough in its treatment, Foresight aims to prepare a man and a woman's body for conception. Counteracting pollution and vitamin and mineral deficiency are at the centre of its work. A whole range of infertility causes have been traced to deficiencies and the presence of toxins in the body. Dietary factors and allergies have been shown to play a part too. For example, current research shows that there could be a link between infertility in women and the presence of Candida albicans in the gut, and that cutting down on sugar and alcohol can help. A survey by Surrey University on the work of Foresight found that around 80 per cent of couples suffering from fertility problems went on to give birth to healthy babies after a Foresight diet and supplement programme.

The basic Foresight plan is a six-month programme which involves eating a wholefood diet, free from additives and organically produced where possible. Natural family planning will be recommended. Mineral analyses will be done and a programme of supplementation and cleansing to rid the body of toxins and balance minerals will be indicated. All water used in drinking and cooking should be filtered. Smoking, alcohol and drugs are prohibited and the use of pesticides in the home, or any food or materials treated with pesticides, should be avoided. Any infections will be treated and tests run for allergies. Sometimes special immunisations, treatments or adapted diets will be advised. You will also have to check with your dentist that he or she isn't still using mercury amalgam.

Remarkable results have been achieved for women and men suffering from unexplained infertility and for women suffering from ovulation problems. Foresight produces informative and helpful booklets, pamphlets and videos. It can help a man and a woman get fit for pregnancy and increase the odds of conceiving all round. You can join Foresight by sending a SAE and membership fee to the address listed in the resources section.

The majority of infertility patients who use alternative therapies use them alongside mainstream treatment. Such therapies help them balance out the system, become stronger, more relaxed and help them feel more positive and in control of their bodies. In short, complementary therapies help them cope with the highs and lows of fertility treatments.

## The highs and lows of fertility treatments

Fertility treatment can take over a woman's life, interfere with her relationships, her sex life, and her work. The whole invasive process can be exhausting and expensive. Many women say that they find their world shrinks. Timing is crucial if the drugs and treatments are to stand a chance of success. All their attention becomes focused upon it. No one obsesses more about motherhood than a woman who wants to have a baby and can't. She begins to think of herself as a baby-making machine.

According to Alice D. Domar, director of the Women's Health Program at the New England Deaconess Hospital, Harvard Medical

School women trying unsuccessfully to get pregnant have stress levels, in terms of anxiety and depression, equivalent to women with cancer, HIV and heart disease.

> Conception can become the main focus of their lives, putting them on a roller-coaster ride that ends with dashed hopes once a month when their periods arrive. Tracking ovulation can take all the fun and spontaneity out of sex and marital disruption is common. (Joan Borysenko, *A Woman's Book of Life* (Riverhead, New York, 1996), p. 196)

But many women are prepared to risk everything if there is even a tiny chance of becoming a mother. There is always the hope that maybe next time the treatment will work. For some, like Ruth and Nicola, it does.

> 'It took me five long years to get pregnant. We tried for two years before seeing a doctor. We had test after test and everything seemed normal. Still no pregnancy. Four attempts at intrauterine inseminations, with Clomid to boost ovulation, also failed. Even more heartbreaking, on one of my Clomid cycles I got pregnant but miscarried. I really lost it then. It took a lot of courage to start another cycle, this time on Pergonal. Thank goodness I did because it was on that cycle that I finally conceived my son.' (Ruth, age thirty-six)

> 'I was told I would never have children after losing both my fallopian tubes to pelvic inflammatory disease and an ectopic pregnancy. I was thirty-six and the NHS refused to treat me because of my age. Even if I had been younger I would still have had to wait years on the waiting list. Steven and I decided to take a risk and pay for private treatment. We were very lucky. I got pregnant with twins. I know lots of people who have several courses of IVF and don't get pregnant, but it worked for us the first time. Life's wonderful now.' (Nicola, age thirty-seven)

But for others, like Tracey, age forty-three, even the most expensive and advanced fertility treatments fail to yield results.

> 'I know that lots of women delay having children because the timing just isn't right. But you have to be aware that the longer you leave it, the less chance you have. That's what happened to me. I thought I could wait until I was approaching forty to have a baby, but I ended up losing baby after baby. Taking time to conceive again. Spending money I didn't have on fertility treatments. Nothing worked. The whole business nearly destroyed me. I can't describe to you the anguish I felt with each miscarriage. Each one was like a bereavement. The day I finally decided to stop putting myself through all the torment was the bleakest, blackest day of my life.'

Infertility is a potential minefield and for a woman and her partner, if she has one, it can erupt at any time. Failing to conceive a baby is devastating. Each month will be the agony of hoping, waiting and grief. This is particularly so for the older woman.

But as well as biological-clock panic, there is also the stress of scheduled sex, and the physical and emotional upheaval of hormonal treatments when the whole body has to be retuned. Huge mood swings and a feeling of being in the grip of forces beyond your control are likely.

Harder still is the constant waiting. Waiting for appointments, tests, referrals, more tests, results of more tests, more tests, results of more tests, an operation, results of an operation, funding decisions, starting treatment, ovulation, results, pregnancy, more tests . . .

There is also the decision of how far you are prepared to go with the treatment. If you are an NHS patient, funding will, of course, determine the limit. If you pay yourself, most clinics recommend four attempts at IVF. If you want to keep trying and trying, much depends on how much your body can endure, but it is important that a woman knows when to stop trying – that she remembers she does have a choice. If she is in a couple, communication is vital – both must agree on what the limit is. If she is single it is vital that she knows exactly how far she is prepared to go.

Coping with the anxiety, stress and uncertainty of infertility treatment is exhausting and stressful. If you have postponed the baby decision it is easy to blame yourself for not trying to have a baby sooner when you were more fertile. It is important that you remember the reasons why you delayed at the time. Your life circumstances were different then and the choice you made was right for you at the time. Don't ever forget that.

You may be able to handle the stress of infertility alone but whether undergoing investigation, contemplating treatment, grieving or considering adoption, at whatever stage, it is worth thinking about the various support groups that exist locally or nationally. The patient support groups run by fertility clinics are particularly helpful (see resources section).

## Coming to terms with infertility

The experience of infertility can be devastating. It is vitally important that a woman suffering from infertility finds some meaning in her experience and comes to terms with her feelings of grief and failure.

A crucial turning point for infertile women is when they begin to see just how much pain might be taken away if they could stop defining themselves as women in terms of their relationship with motherhood. According to Mardy Ireland, if a woman can't accomplish this shift in perspective, she remains in a state of 'pathological grief'. 'Under these circumstances she will continue to feel her feminine self is damaged and that her life is somehow less. She will be unable to identify herself as a woman who is not a mother but still a woman, with a full, satisfying life' (*Reconceiving Women* (Guildford, New York, 1993), p. 40).

Many of us seem to be able to separate female identity from motherhood in our twenties and thirties, but the boundaries sometimes become blurred in later years. A woman who can't have her own baby needs to rediscover for herself that every other option a woman takes is not a poor substitute for motherhood. She needs to understand that being child-free is not an empty, worthless, meaningless way of life, but one that can be full of potential.

Coping with the loss, pain and trauma of infertility is a huge challenge. If healing is to occur it is going to have to centre around grieving for the loss of a child you never had and then reassessing priorities and goals in life. The grieving process is important. Without some kind of emotional expression of grief, feelings remain unresolved and potentially destructive. Counselling may help a couple or a woman through the grieving and healing process.

Healing will focus on taking charge of one's life again. It will likely come about as you learn to separate motherhood from female identity and understand that mothering does not always require children. You can nurture other people, animals, plants, projects, yourself. You can, if you want, be an important influence in a child's life. So many children need the support, interest and care of loving adults. It's what happens after conception that matters the most.

Some women who can't have children do find that once they have been through the intense pain, although the grief never fully goes away, there are ways to transform it. They discover that they can make the most of the hand that fate has dealt them. They find other goals and aspirations. They may even discover that they really are fine without children.

> 'I started trying for children when I was thirty-three. Up until then I didn't really feel ready. Then my husband got a fantastic opportunity to go and live and work in Australia. It meant giving up my teaching job, but I thought it would be a great opportunity to have children and see the world. I tried to get pregnant for a few months but nothing happened. Then we moved to Australia and I went back on the pill for six months during the upheaval. Once we had settled down I again stopped the pill. When there was no pregnancy after six months I sought medical advice and discovered I had a damaged fallopian tube and mild polycystic ovary syndrome. Five years of

unsuccessful fertility treatments followed. It was a heart-breaking, disappointing time.

'During that time I started teaching on a part-time basis at university. I was offered a lectureship at just the same time my doctor finally suggested that we try IVF. I really wanted the job. I knew how stressful, expensive and difficult IVF can be. My husband and I had got used to things the way they were. We had to do a lot of soul-searching. How much did we really want children? Eventually we had to make a decision. It was a difficult and tearful decision, but we decided not to go ahead. We realised that we really could survive without children. That we were doing fine without them.' (Amanda, age forty)

Hearing the voices of women who want to have babies and can't, and gaining an insight into their world of regret, can be particularly distressing if you are undecided about whether or not to have children. Fear of infertility can easily cloud judgment. 'Can I have a child?' may become more crucial than 'Do I really want a child?' but pressure from the biological clock is not a convincing motivation to have a child.

Before you make one of the most important decisions of your life, it is important that you think carefully about all the issues. The next chapter gives advice about how you can maximise your chances of fertility while you ponder your options and the clock ticks.

# 11

# Healthy Tips while the Clock Ticks

*Today's woman wants to know how her body works. She deserves in-depth information about how her reproductive system functions and how she can protect her health and enhance her fertility.*

(Robert Franklin and Dorothy Brockman, *In Pursuit of Fertility* (Henry Holt, New York, 1995), p. ix)

There are no guarantees as far as fertility is concerned, but during the decision-making process there are things you can do to optimise your chances of fertility, and reduce the chances of miscarriage should you decide you want to have a baby.

If you are worried about how age will compromise your fertility, a healthy lifestyle can help balance the effect of added years. Good health will not only maximise your chances of fertility, but it will also help you cope with biological-clock anxiety.

Simple leaflets on good health are readily available from doctors and chemists. In this chapter we'll talk about staying healthy while the biological clock ticks – everything from a healthy diet to exercise and stress reduction.

## A healthy diet

Today's woman, by the time she is in her mid thirties, has likely put a large strain on her digestive system. She has probably consumed a high percentage of refined carbohydrate, dairy foods, caffeine and alcohol. Diets high in sugar and alcohol can promote the growth of Candida albicans, a yeast-like organism which causes thrush and poor absorption. She may have taken the pill, which can affect the way food is absorbed, and been on regular courses of antibiotics, which can also contribute to poor digestion.

Poor digestion usually combines with a build-up of toxins because the body is less efficient at flushing them out. Attention to nutrition, if there are problems with poor digestion, is essential because it is often hard to get the necessary levels of vitamins and minerals from the food we eat owing to the way modern food is grown, stored, cooked and transported.

Research is proving that a high percentage of women diagnosed with infertility do have nutrient deficiency due to poor digestion. Without enough zinc, vitamin A and magnesium, for instance, your sex hormones can't even be produced, let alone function. These substances are needed at every stage of the reproductive process.

For optimum fertility a healthy, nutritious diet is perhaps the single most important factor. A healthy, balanced diet should include all the essential nutrients: carbohydrates, fats, proteins, vitamins, minerals and water. Junk food, caffeine, alcohol and smoking should be minimised if they can't be avoided. Junk food is high in chemical additives. Smoking, caffeine and alcohol harm male and female fertility.

For women in their thirties and forties concerned about hormonal imbalance and declining fertility, nutritionist Louise Gittleman recommends a diet with a balance of 40 per cent carbohydrates, 30 per cent fat and 30 per cent protein: 'For hormonal regulation and sustained physical and mental effort, nothing beats a diet with a 40/30/30 ratio' (*Before the Change: Taking Charge of Your Perimenopause* (HarperSanFrancisco, 1998), p. 15).

Carbohydrates are an important source of energy, vitamins and fibre. There are two kinds of carbohydrates, simple and complex. Simple carbohydrates, like white flour, pasta and white sugar, are absorbed very quickly by the bloodstream and are less desirable than complex carbohydrates, like fruit, vegetables and oatmeal, which are digested slowly and release energy steadily. Complex carbohydrates also tend to be higher in fibre, or roughage, which is important for hormone regulation and healthy digestion. If you eat as much unrefined grain and fresh fruit and vegetables as possible you can get your daily supply and keep regular.

Eating protein-rich foods, like lean meat, fish, soy, eggs and cheese, is also important. Proteins perform many life-giving functions inside our bodies and are essential for hormone production.

Many weight-conscious women in their thirties tend to eat low-protein, low-fat and high-carbohydrate diets. But to optimise our chances of fertility our bodies need a balance of all the essential nutrients.

Fats are an important part of our diet. Dietary fats provide the essential fatty acids required to produce hormones. If we don't eat enough fat, hormone production will be compromised. A diet too low in fat can cause fertility problems. A weight-obsessed society has indoctrinated us against fats, and many of us don't appreciate how important the right kind of fat is. It is true that some fats are high in calories and increase the risk of infertility and artery-clogging cholesterol. These tend to be saturated fats from red meats and full-fat dairy products, and they should be avoided. But unsaturated fats, such as those found in seafood, soya beans, leafy vegetables and lean meats have many health benefits.

Especially important for women in their reproductive years are the essential fatty acids. For hormonal regulation we need a diet which is rich in omega 3 and 6 food sources. Flaxseed oil, sea vegetables, green vegetables and cold-water fish are rich sources of omega 3 essential fatty acid. Unrefined vegetable oils, like sunflower oil, soy oil, olive oil, as well as green leafy vegetables, liver and kidneys are rich food sources of omega 6.

Our bodies are made up of about two-thirds water, so the intake and distribution of fluid is important for hormonal regulation. If the body is deprived of water, blood volume is reduced and the blood won't circulate to the reproductive organs so effectively. The solution is to drink enough water – six to eight glasses a day is recommended. If you find it hard to drink all that water, try adding a slice of lemon or some other flavouring to the water.

Vitamins and minerals are essential because they help various bodily processes and chemical reactions take place. As pollution increases your need for nutrients, it might be wise to ensure that you are getting all the vitamins and minerals your body needs. (One way to ensure this is to eat food as fresh as possible as over-cooking and processing can kill essential nutrients.)

Here's a brief look at the major vitamins and minerals every woman postponing the baby decision should ensure her diet includes.

## Vitamins

- Vitamin A fights infection and prevents dry skin and poor bone growth. It is found in vegetables, milk, butter, margarine and egg yolks.
- Thiamin is important for carbon dioxide removal during respiration. It is found in wholegrain nuts and seeds.
- Riboflavin is needed for cell renewal and is found in milk, meat, eggs and leafy vegetables.
- Niacin helps to prevent disease, improve the mood and promote a glowing complexion. You can find it in milk, eggs, cheeses and fish.
- Pantothenic acid is essential for energy metabolism and stress reduction. It is found in many foods such as meat, fish, poultry, wholegrain cereals and dried beans.
- Vitamin B6 is important for the metabolism of proteins. It can relieve water retention, bloating, skin problems, depression and symptoms of perimenopause. It has the ability to boost progesterone levels and reduce oestrogen levels. It is also needed for the normal secretion of serotonin, the brain neurotransmitter that regulates mood, sleep and appetite. Vitamin B6 is also important in collagen formation and bone strength. Women on the

contraceptive pill must ensure that they get adequate vitamin B6. Food sources include spinach, bananas, liver, nuts, meat, fish, legumes and green peppers. Vitamin B6 can be taken on its own but the B vitamins are absorbed better if taken in complex form.

- Vitamin B12 promotes healthy skin and helps maintain a healthy nervous system. It is found in meat, milk, eggs, cheese and fish.
- Biotin is important for carbohydrate and fat metabolism. It is found in liver, peanuts and cheese.
- Folate promotes cell production and healthy skin and is found in leafy green vegetables, chicken, liver and kidneys.
- Vitamin C, found in green vegetables and citrus fruits, prevents colds, heals wounds and is essential for normal metabolism and the reduction of menstrual problems. It boosts the immune system, fights fatigue and protects us from toxins.
- Vitamin D, good for bone growth and calcium absorption, is also an aid in relieving menstrual problems and is found in tuna fish, eggs, butter and cheese.
- Vitamin E promotes blood clotting and is found in milk, vegetables, liver, rice and bran. Diets low in fat are often deficient in vitamin E which is essential for hormonal balance in women. Good food sources include vegetable oils, nuts and seeds. Vitamin E can alleviate perimenopausal symptoms such as hot flushes, breast tenderness.
- Vitamin K is active in maintaining the involuntary nervous system, vascular system and involuntary muscles. It is found in wheatgerm, vegetable oil and wholegrain bread and cereal.

## Minerals

- Calcium prevents blood clotting and is necessary for bone growth, healthy teeth and iron absorption. It is found in milk and milk products, egg yolks, green vegetables and shellfish. Calcium deficiency causes bone loss as the milk ads constantly remind us. Most of us know that calcium deficiency is a health risk. What we don't know, however, is that most of us aren't calcium deficient; we simply lack the ability to utilise the calcium we have because we don't have enough magnesium in our diet. Calcium needs magnesium to be incorporated into your bones. If there is not enough magnesium, the extra calcium will collect in soft tissues not bones, and cause arthritis. Dairy products contain more calcium than magnesium. Many women in their thirties have a calcium–magnesium imbalance because they avoid magnesium-rich foods like almonds, nuts, seeds, milk, grains, vegetables, fruits, cereals and sea vegetables. Magnesium deficiency can cause peri-menopausal symptoms. It has a sedative effect, and symptoms include depression, nervousness, anxiety, concentration problems, frequent urination, body odour increases, constipation and fatigue.

- Chromium normalises insulin levels so that sugar can be processed more quickly. Imbalances in the blood-sugar hormones are interconnected with imbalances in the sex and stress hormones. Foods rich in the mineral include brewer's yeast, liver, some shellfish and mushrooms.
- Copper is an aid in the metabolism of iron and is found in liver and whole grains.
- Fluoride strengthens teeth and is found in fluorinated water and tea.
- Iron is the basic component of blood haemoglobin and prevents anaemia. If your diet is deficient in iron you will feel tired and lethargic and listless. You may also have pale skin, dark circles under the eyes and feel the cold more. Irritability and headaches are also associated with iron deficiency. A lot of physical activity, pregnancy and nursing can severely deplete iron amounts. As well as this, iron deficiency can cause eating disorders and interfere with thyroid function. On the other hand too much iron is not ideal either. This can increase the risk of heart disease and cancer. Women who don't menstruate or who have stopped menstruating often have high levels of iron. Iron can be found in dark green leafy vegetables, beans, eggs, dried fruit and liver. Iron intake increases if vitamin C intake increases. Red wine and dark beer also increase iron absorption. On the other hand caffeine, aspirin and some food preservatives interfere with iron absorption.

- Iodine is an aid in regulating energy use in the body. It is found in seaweed and seafood.
- Potassium is needed for healthy nerves and muscles and is found in milk, vegetables and fruit.
- Sodium helps maintain adequate water in cells in the body and is found in table salt, milk and meat.
- Phosphorus, found in milk, yoghurt, yeast and wheatgerm, is required for bone growth, strong teeth and energy transformation.
- Zinc plays an essential part in keeping the reproductive organs and the body's enzyme systems healthy. Zinc is also important for perimenopausal women because it is needed for bone formation. Zinc supplements can actually slow bone loss and boost the immune system. Women who are zinc deficient tend to have high copper levels. Low zinc and high copper levels will cause perimenopausal symptoms. Zinc deficiency is caused by stress and a high sugar, carbohydrate diet. Certain medications can cause zinc deficiency as can alcohol and high-fibre vegetarian diets. Food sources of zinc include red meat, egg yolk, milk, nuts, peas, beans, seafood and whole grains. Vegetarians should certainly ensure that they get sufficient zinc intake.

## Taking supplements

It can be confusing to know how to combine all these vital ingredients to ensure health and fertility. Finding that balance is complicated even further by the fact that there are two schools of thought. Some experts believe that we can get all we need from the food we eat – that a healthy diet is all that is required. Other experts say that our busy lifestyle and the way food is produced today make it virtually impossible to get all the nutrients we need.

Every woman has to rely on her own judgment here, but it is perhaps safer if you are concerned about fertility declining with age to take supplements. Arguably the most important supplement you should be taking is folic acid. Every woman thinking about having children should ensure she takes a folic acid supplement of 0.4

milligrams every day. Folic acid deficiency is associated with neural tube defects in babies-to-be. As well as watching your folic acid, pay particular attention to intake of iron and vitamin D, as the requirements for these particular nutrients increase dramatically should you get pregnant.

There are so many things you can't control about your fertility, but diet is at least not one of them. Research has shown that remarkable improvements in fertility can result from large doses of fertility-enhancing vitamins and minerals listed below. However, always consult a nutritionist because self-prescription can be dangerous and some nutrients (vitamin A and selenium, for instance) are dangerous if you take too much.

## Vitamin and mineral supplements

- Vitamin A, found in eggs, butter, green fruit and vegetables. Too much vitamin A can be harmful, and it is safer to take betacarotene. Betacarotene is unlike the vitamin A from animal products, and is water soluble so it cannot build up in the body. Foods rich in betacarotene include spinach, broccoli, red and yellow peppers.
- B vitamins are found in wheatgerm, wholegrain products, yeast extract and brown rice. They are crucial for women drinking large amounts of coffee. Most important are: vitamin B2, found in yoghurt, yeast, eggs, green leafy vegetables, mushrooms, fruits and cereals; vitamin B3, found in lean meat, fish, peanuts, bran and beans; and vitamin B12, found in fish, meat, eggs, milk and cheese. Vitamin B12 is especially important for vegans and vegetarians. It is advisable to take B vitamins in complex form rather than individually.
- Folic acid, found in spinach, green beans, brussels sprouts, milk and fruit.
- Vitamin C, found in raw fruit and vegetables.
- Vitamin E, found in wholegrains, vegetable oils, seeds, nuts and green leafy vegetables.
- Calcium, found in dairy products, fish, nuts and dried fruits.

- Essential fatty acids, found in evening primrose oil, olive oil, flaxseed oil, oily fish, avocados, sunflower and sesame seeds.
- Iron, found in eggs, fish, dried milk and dark green vegetables.
- Manganese, found in wholegrains, nuts, onions.
- Magnesium, found in milk, nuts, seafood, wholegrains.
- Selenium, found in herring, tuna, wholewheat, broccoli, garlic.
- Zinc, found in seafood, milk, wholegrains and dried fruit.
- Antioxidants: a number of vitamins and minerals are called antioxidants. Antioxidants can counter the negative effects of free radicals which are molecular fragments that can cause damage to our fertility and our bodies. Ensuring adequate intake of the antioxidant vitamins and minerals can be extremely beneficial. The most important food-derived antioxidants are betacarotene (the non-toxic part of vitamin A), vitamins C and E as well as certain B compounds and the minerals selenium, manganese, copper and zinc.

## Preventable risks

It is sensible to tackle dependency on anything known to inhibit fertility, like alcohol, drugs, smoking and caffeine.

### Alcohol

Studies show that women are drinking more than ever before, especially the young and career-minded. Any woman who drinks heavily reduces her chances of fertility and doubles her risk of miscarriage and having a baby with a deformity. The risks starts to rise at the equivalent of four measures of spirits or glasses of wine, or two pints of beer and cider per day, increasing steeply at about three times these amounts. Alcohol-related risk is greater if you are over the age of thirty-five, use the contraceptive pill, smoke and your diet is poor.

## Drugs

Drugs alter the chemistry of a woman's body, and you should always check with a medical expert before taking any form of medication. This includes over-the-counter remedies and alternative therapies. Street drugs, like ecstasy, cocaine and heroin, are known to reduce fertility and increase the risk of miscarriage. If you have a condition that requires medication, tell your doctor that you are concerned about your fertility so that you know all the options available to you.

## Smoking

Smoking is a well known inhibitor of fertility. Smoking during pregnancy increases the chances of miscarriage, premature birth and low birth-weight babies. If a woman is over thirty-five smoking also increases the risk of babies born with birth defects. Many women smoke because they think it keeps their weight down, but it is possible to stop smoking and not to put on weight. Books on quitting smoking can offer support and ideas if you are addicted. Or you may prefer to ask your doctor's advice.

## Caffeine

Too much caffeine may also prove to be harmful because caffeine depletes the body of essential nutrients. Try to limit the amount of tea, cola, coffee and chocolate you consume to a sensible amount.

### Tips for a healthy diet

- Drink plenty of mineral and filtered water.
- Reduce animal fats.
- Cut down on alcohol.
- Cut down on caffeine.
- Cut down on refined carbohydrates and sugar intake.
- Ensure that enough essential fatty acids are consumed.

- Eat more fibre.
- Take a daily vitamin and mineral supplement.
- Eat food as fresh as possible, rather than processed, frozen or tinned.
- 'Stir fry', grill, roast or stew rather than deep fry.
- Steam rather than boil.
- Cook with oils such as olive and sunflower rather than butter or lard.
- Eat plenty of raw fruit and vegetables.

## Weight management

Weight, in particular body fat, has a direct influence on egg production and can be the cause of a woman's inability to conceive. Women who are too fat or too thin can experience fertility problems.

Excess weight can affect ovulation (that time of the month when ovaries release an egg). Fat cells produce oestrogen, and too much prevents the egg from being released. Oestrogen is the main ingredient of some contraceptive pills. Obesity can also affect insulin levels in women, which can cause ovaries to overproduce male hormones and stop releasing eggs.

Low weight or weight loss can lead to a temporary weakening of an important hormonal message that the brain sends the ovaries. In some cases eggs may be produced and released, but the lining of the uterus is not ready to receive the fertilised egg. In other cases ovulation does not occur.

Eating a healthy, balanced diet should ensure that a woman stays at a healthy body weight. It should also regularise any weight problems over time. Erratic eating habits, dieting and weight fluctuations can all damage fertility. After years of yo-yo dieting and weight gain and loss, some women do suddenly experience problems with their fertility.

If you are postponing the baby decision and constantly dieting to maintain a low body weight, it is vital that you ensure your body is getting all the nutrients it needs to maintain ovulation and healthy reproductive function. Women who suffer or have suffered from severe eating disorders may find that their obsession with body weight has damaged their fertility.

Weight has a way of creeping up over the years as you get older, but you will conceive more easily and be less likely to suffer problems like high blood pressure or diabetes during pregnancy if you are within the normal range.

The body mass index is an accepted standard based on the relationship between a person's weight and the square of their height, regardless of age or build. A BMI between 20 and 25 is ideal for optimum health. A little higher or lower is fine, but lower than 17 or higher than 30 is cause for concern (see Figure One).

If you need to lose weight, aim for safe, steady weight loss of one or two pounds a week and increase the amount of exercise you take. Choosing foods high in nutrients and low in calories will be important. You might want to ask your doctor's advice or to see a nutritionist. If you need to gain weight you may need to reduce the amount of exercise you do and increase the amount of nutritious food you eat to let your weight rise a little.

It is also important to check if you are an 'apple' or a 'pear' (see Figure Two). If your waist measurement divided by your hip measurement is 0.85 or less you are a pear; 'apples' have a waist to hip ratio over 0.85. 'Apples' put on fat around the stomach and there are health risks attached to this like heart disease. 'Pears' gain the weight around the hips and thighs and this seems to be safer. 'Pears' also tend to conceive more easily than 'apples'.

## Avoiding pollution

Avoiding pollution and harmful radiation may also be beneficial. Unfortunately in this modern world it is difficult to avoid. Metals and chemicals are found in a wide range of items including white tooth fillings, detergents, pesticides and plastic packaging and drinking water. Some metals, particularly lead and cadmium, have

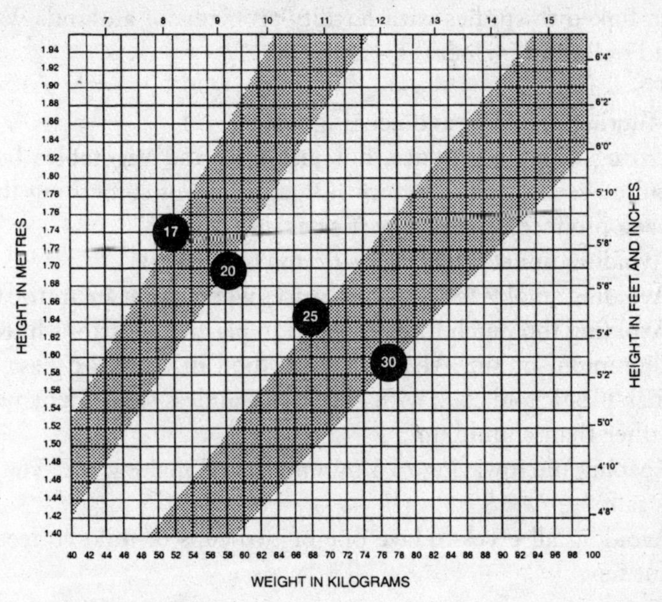


**Figure One**

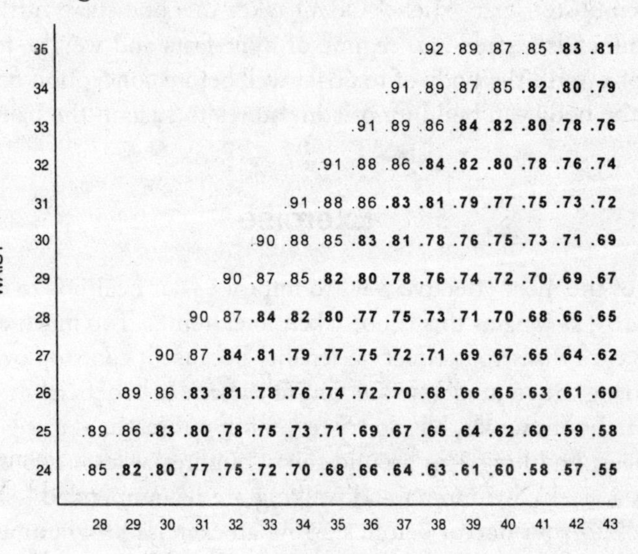

| WAIST \ HIP | 28 | 29 | 30 | 31 | 32 | 33 | 34 | 35 | 36 | 37 | 38 | 39 | 40 | 41 | 42 | 43 |
|---|---|---|---|---|---|---|---|---|---|---|---|---|---|---|---|---|
| 35 | | | | | | | | | | | .92 | .89 | .87 | .85 | **.83** | **.81** |
| 34 | | | | | | | | | | .91 | .89 | .87 | .85 | **.82** | **.80** | **.79** |
| 33 | | | | | | | | | .91 | .89 | .86 | **.84** | **.82** | **.80** | **.78** | **.76** |
| 32 | | | | | | | | .91 | .88 | .86 | **.84** | **.82** | **.80** | **.78** | **.76** | **.74** |
| 31 | | | | | | | .91 | .88 | .86 | **.83** | **.81** | **.79** | **.77** | **.75** | **.73** | **.72** |
| 30 | | | | | | .90 | .88 | .85 | **.83** | **.81** | **.78** | **.76** | **.75** | **.73** | **.71** | **.69** |
| 29 | | | | | .90 | .87 | .85 | **.82** | **.80** | **.78** | **.76** | **.74** | **.72** | **.70** | **.69** | **.67** |
| 28 | | | | .90 | .87 | **.84** | **.82** | **.80** | **.77** | **.75** | **.73** | **.71** | **.70** | **.68** | **.66** | **.65** |
| 27 | | | .90 | .87 | **.84** | **.81** | **.79** | **.77** | **.75** | **.72** | **.71** | **.69** | **.67** | **.65** | **.64** | **.62** |
| 26 | | .89 | .86 | **.83** | **.81** | **.78** | **.76** | **.74** | **.72** | **.70** | **.68** | **.66** | **.65** | **.63** | **.61** | **.60** |
| 25 | .89 | .86 | **.83** | **.80** | **.78** | **.75** | **.73** | **.71** | **.69** | **.67** | **.65** | **.64** | **.62** | **.60** | **.59** | **.58** |
| 24 | .85 | **.82** | **.80** | **.77** | **.75** | **.72** | **.70** | **.68** | **.66** | **.64** | **.63** | **.61** | **.60** | **.58** | **.57** | **.55** |

**Figure Two**

been linked to studies with fertility problems of all kinds. Ways to avoid pollution include:

- Filtering drinking and cooking water.
- Eating organic produce, not just fruit and vegetables but also pulses, seeds, dairy produce. 'Organic' means something that has been produced without pesticides and drugs.
- Avoiding unnecessary exposure to chemicals.
- Avoiding smoky atmospheres, places where there are traffic fumes.
- Avoiding too much food wrapped in plastic, in cartons lined with aluminium or tin. At home store food in china or glass rather than plastic and cook with cast iron, stainless steel or enamel pans rather than aluminium.
- Keeping microwaving to a minimum and making sure your oven is safe.
- Avoiding all excess. Have one or two cups of mild coffee a day, not ten.

French obstetrician Michel Odent takes this one stage further and recommends a strenuous regime of mini-fasts and weight-recovery sessions to rid the body of toxic fat well before conception occurs so that the body can build up pristine stores to sustain the baby.

## Exercise

One of the most effective ways to improve your health is to exercise regularly, as long as this is not taken to extremes. Too much exercise can have a disastrous effect on fertility because it can stop ovulation occurring. But a moderate amount of exercise is beneficial. It doesn't have to be strenuous, but do try to get out of breath, as it is good for the heart and lungs, for around thirty minutes at least two or three times a week. Swimming and walking are recommended.

Talk to your doctor before starting an exercise programme if you have high blood pressure, you smoke or your BMI is over 30.

Exercise helps maintain reproductive health and the hormonal system. Exercise keeps our heart and lungs in good condition, promotes energy and health, removes toxic substances from the body, increases calorie expenditure and aids weight management, and improves posture and self-image. It is also helpful in reducing stress, anxiety and depression and in dealing with psychological problems. Exercise releases endorphins, which can improve mood. Even a thirty-minute walk round the block can reduce anxiety.

## Stress reduction

More and more these days we are hearing about celebrities or career women who are slowing down the pace of their lives to try and have a baby. It's impossible to generalise, but it does seem that there is a connection between infertility and a demanding, hectic lifestyle with no time for relaxation.

Stress reduction is important for optimal health. Stress occurs when there is an imbalance between the demands of your life and your ability to cope with these demands. Constant stress over a long period of time will upset hormonal balance, deplete the body's resources, cause anxiety and depression, diminish the immune system and inhibit fertility.

'A certain amount of stress is invigorating – without any we would be dead – but excessive stress, especially if it exceeds our ability to cope, can be harmful' (Robert Franklin and Dorothy Brockman, *In Pursuit of Fertility* (Henry Holt, New York, 1995), p. 117). We need a certain amount of stress and challenge in our life, but it should be a positive kind of stress, not stress that damages our health. Negative stress should be avoided, or if it can't be avoided we need to learn how to deal with it in a positive way. In the words of Joan Borysenko, it depends on our constitutions:

> Most of us will feel that life is out of control in some way. Whether we see this as a temporary situation whose resolution will add to our store of knowledge and experience, or as one more threat demonstrating life's dangers is the most crucial question both for

the quality of our life and our physical health. (Joan Borysenko, *A Woman's Book of Life* (Riverhead, New York, 1996))

One way to cope better with the stresses in our lives is to ensure that we are in good health. Eating right and exercising regularly are essential for good health. Getting enough sleep is also beneficial. Sleep is biologically necessary. Most of us need around eight hours a day and if we do not get it our bodies and minds don't function so well.

Relaxation is an important stress management technique. If you find it hard to relax you might like to try listening to relaxing music or even a relaxation tape. Alternative therapies can also be helpful for relaxation, good health and fertility. Massage with aromatherapy oils, for example. Oils need to be diluted for massage and those thought to be good for fertility include chamomile, frankincense, geranium, lavender, rose, ylang ylang.

Techniques like yoga or tai chi can have astonishing effects on women who are stressed. Deep slow breathing through the nose rather than the mouth and into the stomach, not the chest, can calm both body and mind and help us cope with stress. Simple yoga breathing exercises, for example, breathing in slowly through the nose while counting to five in your head, holding your breath for a count of five, breathing out slowly through the nose for a count of five, waiting a count of five and repeating as often as you like, may also help. Concentrating on breathing and counting can be wonderfully calming for your mind while the regular breathing will calm the body.

There are many delightful ways to relax. Soothing music, soaking in a hot tub, laughing more, interacting with others, helping others, having a positive outlook, cultivating outside interests and diversions from your routine. You should not think of it as time lost, but as time gained. When you return to your daily routine you should feel refreshed, energised and more in control.

## Fertility awareness

While you are making your decision about parenting it is important to know your body better and in particular your ovulation cycle. Understanding the ebb and flow of your hormones as they regulate your menstrual cycle can be particularly helpful.

The menstrual cycle is run by three glands: the hypothalamus in the brain, the pituitary in the brain and the ovaries. These produce hormones, the chemical messengers of the brain.

At the beginning of a menstrual cycle the hypothalamus hormone, gonadotrophin-releasing hormone (GNRH) stimulates the pituitary gland to secrete follicle-stimulating hormone (FSH) until about day twelve of a typical 28-day cycle. FSH stimulates the follicles inside the ovary to develop and this first phase of the cycle is called the follicular phase. FSH also stimulates the production of oestrogen. Rising oestrogen levels act on the pituitary to cause a surge of luteinising hormone (LH) which triggers ovulation at the half-way point of the cycle around day fourteen.

In the ovary the empty follicle, after releasing the egg in ovulation, becomes a special structure and starts to produce progesterone hormone which makes the lining of the womb or endometrium thicken ready for a fertilised egg. This empty follicle is called the corpus luteum and the second half of the cycle is called the luteal phase. If pregnancy occurs, progesterone will support the pregnancy until the placenta forms. If it doesn't occur, progesterone levels fall around fourteen days after ovulation and menstruation occurs.

If periods are consistently irregular, absent, too long or heavy, this is a sign that there is some problem with the normal hormonal cycle and potential problems with fertility. The right hormones are not being released, or they are being released at the wrong time. Ovulation probably isn't happening and conception can't occur.

Many women aren't really sure what an irregular period is. Although there are always exceptions, in general a normal range in cycle length for a woman in her twenties is twenty-two to thirty-nine days, in her thirties and forties it is twenty-three to thirty-five days. The length of a cycle refers to the number of days from the first day of bleeding of one cycle to the first day of bleeding in the next. The average is twenty-eight or twenty-nine days. The length of a period is the number of days in a single cycle bleeding occurs, usually five. Cycles typically tend to shorten by a few days the older a woman gets, but as menopause approaches they get longer.

Periods rarely arrive on the dot and each month there will be variations. For example, a cycle may be twenty-eight days one month and thirty-three the next, and this is nothing to worry about. Lifestyle changes such as a change in diet, weight loss, stress overload, travel or excessive exercise can also throw off an otherwise normal cycle, causing a woman to have a later period, to skip a period or even to menstruate twice a month. This isn't anything to worry about if it happens occasionally. However, if there are period irregularities for more than two or three months when periods are less than twenty-two days apart or more than thirty-five days apart it is likely that ovulation is not occurring regularly.

Not ovulating is a common cause of irregular periods and infertility. So keeping track of your menstrual periods and paying attention to any changes is important. Bear in mind that during perimenopause and menopause, when fertility begins to decline, periods get more and more irregular.

You can tell if you are ovulating and therefore fertile by taking body temperature readings morning and night, on a daily basis. These will show, as do changes in cervical mucus, when the cycle is peaking. At ovulation body temperature drops slightly and then rises by 0.2 to 0.4 °C. It stays high until the next period. Cervical mucus also becomes clear and stretchy resembling egg white. Ovulation predictor kits can also pinpoint ovulation with great accuracy.

## Contraception

The method of contraception you choose to use may affect future fertility.

A woman is fertile for several days around ovulation. Knowing when you are ovulating can therefore be an ideal method of birth control. It is natural and will not compromise fertility in any way. But it is difficult to master and not 100 per cent effective. Many women postponing the baby decision don't want to risk pregnancy before they are ready and prefer to use safer forms of contraception, like the condom, coil and the pill.

The condom is worn by the male on his penis during intercourse to prevent sperm entering the vagina. Used carefully and skilfully this is extremely effective. The coil prevents an egg being implanted in the uterine wall. It carries with it a small risk of infection and ectopic pregnancy, but it is effective.

The pill, which is the preferred choice of contraception for most women in their reproductive years, is around 99 per cent effective. It prevents pregnancy occurring by inhibiting the pituitary gland from releasing hormones which trigger ovulation. Ovulation does not occur. The progesterone-only pill doesn't prevent ovulation, but makes cervical mucus sperm-resistant. It also carries with it a slight risk of ectopic pregnancy.

When the pill first came out there were fears that it would affect women's fertility. Although most doctors consider the pill safe for long-term use, and some studies show only a two-to-three-month delay in normal menstrual cycling after going off the pill, other evidence tells a somewhat different story.

In a study of women aged between thirty and thirty-five it was found that many experienced a much longer delay when trying to conceive after withdrawing from the pill. In about half the cases it took these women a year longer than those who had been using the diaphragm. It was six years before conception rates were the same. One wonders what a study conducted on women over the age of thirty-five would yield. What would be the delaying period then? If a woman is thirty-eight and trying for a baby, waiting two to three years for her body to adjust after taking the pill is taking a risk with her fertility.

The pill also alters vitamin and mineral levels. That's why doctors recommend that a woman gives her body a few months to recover before she tries to conceive, and that she switches from the pill to a barrier method such as a condom. The pill reduces the amount of folic acid, vitamin C, vitamin E and zinc in your blood, while iron and copper levels are raised.

Some experts believe that because the pill depletes your body of essential nutrients it can lead to reduced fertility. Levels of vitamin A can go up when a woman is on the pill, which can be toxic. The pill lowers levels of B1 and B2 in the blood and possibly vitamin B6. Folic acid, vitamin B12 and vitamin C are all affected by the pill. The pill is also thought to increase levels of iron and copper. But zinc is perhaps the most important mineral affected. The pill is thought to change a woman's capacity to absorb zinc, and zinc deficiency has been linked to infertility. If a woman is on the pill, has been on the pill, or is coming off it, a good multivitamin is sensible. A daily folic acid supplement (0.4mg) is especially recommended for women who want to maintain fertility and one day have a healthy baby.

The pill, because it alters the hormonal balance of the body, can also mask conditions, like polycystic ovary and endometriosis, which can affect fertility. A woman may be unaware that she is suffering from a condition which can compromise her fertility because she has regular bleeds on the pill. It is only when she withdraws from the hormonal control of the pill that the condition becomes apparent.

It is not all bad news as far as the pill is concerned though. Some women do experience a peak of fertility when they stop taking it. In short, there really is no telling whether or not a woman will experience a delay in conceiving or whether she will conceive easily. The choice is up to the individual. However, after the age of thirty-five, a woman postponing the baby decision might be wise to consider switching to a more natural form of contraception, considering the possibility of delay should she decide to conceive and the fact that the pill can alter nutrient levels in the body.

## Fertility-boosting planner

- If possible stop using the pill, hormone injections or implants. This will restore your vitamin and mineral levels which are altered by the hormones and let your natural cycle re-establish itself.
- Eat a nutritious, healthy diet.
- Exercise regularly.
- Manage your weight.
- Take a folic acid supplement from your GP or chemist.
- Stop drinking, drug and smoking habits.
- Practise fertility awareness.
- Reduce the amount of stress in your life.
- Regular pelvic examinations and well-woman visits to the doctor are another safeguard against unexpectedly discovering fertility problems later on.

Leading a healthy lifestyle and practising fertility awareness won't ensure fertility, but they will optimise your chances of it and help you cope better with pressure from the biological clock. The final chapter of this book will look at other ways you can help yourself cope with the clock.

# 12

# Coping with the Clock

*The popular concept of the 'baby button' is one of the ways in which women are pressured to have children: the 'well, I ought to have one, just in case I regret it later' philosophy.*

(Jane Bartlett, *Will You Be Mother?*
(Virago, London, 1994), p. 60)

You can't slow down or stop the biological clock, but this doesn't mean you have to be pressured by it. While it may be true that you are working under a deadline it is vital that you don't panic. Panic is not conducive to making a wise choice.

If having a baby is the most important thing in your life at the moment, then stay calm, plan accordingly and set about making it happen. Just because the biological clock is ticking loudly it doesn't mean you have to die childless. The really determined woman will find a way, if not by having babies of her own then via adoption, fostering or caring for children in other ways.

You may meet the right man in time, but if you don't, there is the single motherhood option to consider. Should fertility be an issue, remember that women all over the world are waiting until their late thirties or early forties to reproduce and many of them are conceiving with the aid of fertility treatments.

If you do have problems conceiving don't think of waiting as a mistake. If you had an abortion earlier in life don't blame yourself or feel guilty. Never lose sight of the fact that there were good reasons why you have been delaying the baby decision. Should fertility treatments fail, you can always consider adopting, fostering or mentoring a child.

If you aren't sure you want to have a baby at the moment it's important that you stay calm too. Don't let the biological clock pressure you into a decision you may regret later. Children are not fashionable accessories. Just because a woman can have a baby does not mean that she should. Having a child because your biological clock is ticking and you feel you don't want to miss out is not a good reason to have a child. To avoid making the wrong decision at this crucial time in your life you should think very carefully about why you want to have children.

Try not to get bogged down with details. Thinking about day-care, expenses and so on is important, but for now you need to think about having a baby on a much deeper, emotional level. Your focus needs to be on how much you want to become a mother. How much do you want the experience of parenthood? This isn't the same as asking 'How much do I want to be pregnant?' Some women do want to experience pregnancy, but are not at all excited by the prospect of raising a child.

Ask yourself if you are pushing yourself into motherhood because you fear your fertile years are drawing to a close. Do you really want a child or are you frightened of missing out on what everybody else seems to be doing?

Other questions you need to ask yourself are: would you still feel complete if you reach menopause without a child? Is your desire for a child an attempt to make up for some loss or lack in other areas of your life? The desire for a baby can be strong after the end of a relationship, the loss of a job, the death of a parent, disappointments at work and so on. Ask yourself if you can distinguish between wanting to be a parent and making up for losses in other areas of your life.

Ask yourself what your motivations are to have a child. How

wholesome are they? Do you want to sustain an ailing relationship? Do you want to avoid loneliness in old age? Don't feel guilty if you do have dubious motivations. Acknowledge they exist and try to deal with them before you decide to have a baby.

According to Dr Susan G. Mikesell PhD, psychologist in Washington, DC, consulting psychologist for the Montgomery Fertility Institute in Bethsada, Maryland, and consultant for the Prevention Biological Clock Anxiety webpage (http://www.healthyideas.com/healing/cond_ail/bioclock.html), 'the way you answer these questions tells you different things' about yourself and what your motivations are.

Toni Bernay PhD, a Beverly Hills psychologist who has seen a tremendous rise in the number of patients trying to decide whether to have children, also gives great importance to motivation and finding out where your priorities really lie. Bernay believes that the most important task, if a woman has been postponing children, is to examine the reasons why she doesn't have them already by the time the biological clock starts ticking. 'Often they've been focusing on their careers. That's the first layer. When you lift that layer up, there are often other issues, such as not wanting to be in a committed relationship or not wanting the obligations or responsibility children represent' ('Beating the Biological Clock With Zest', American Psychological Association webpage – http://www.apa.org/monitor/feb96/fam40a.html, p. 1).

Examining the reasons you have been delaying the baby decision can help you understand what your priorities in life are. And now, with the biological clock ticking, you can reassess the situation and determine how high on this list of priorities having a baby is. Is this really the most important thing in your life right now?

Sensitivity to the issues, self-awareness and information are the key factors which can make the choice easier, but don't let them take over your life. Try if you can to limit the amount of time you think about why you want children to short periods. Set aside daily sessions of twenty or thirty minutes. During this time reflect on the issues involved, but when the time is up stop thinking about them and get on with the rest of your life. This may sound slightly ridiculous, but it really can help you contain and control the anxiety and prevent you rushing into a decision out of panic. And giving yourself these focused sessions, where you analyse the issues over and over again, will help the decision you finally make crystallise.

## Advice, therapy and support

With postponement of motherhood becoming the norm – the average age for first-time mums is now edging towards thirty – biological-clock anxiety looks set to become the latest contemporary neurosis. Doctors are only just beginning to appreciate the extent to which it can destroy quality of life and are recommending therapy, support and counselling.

Worrying about whether you can or should have a child can take over your life if you let it. If fears about declining fertility, and anxiety about whether or not to have a baby, reach a point that they are interfering with other aspects of your life, your job, your relationships, it is important that you visit your doctor or seek advice, therapy and support.

A woman may also benefit from therapy and outside support if she is torn apart because her partner doesn't want a child and she does, if she is thinking of getting pregnant even though her partner objects, if she is single and anxious to have a baby, and if she is extremely depressed by unsuccessful attempts at pregnancy.

Some women who think they want to have children are confused about the source of their feelings and find therapy helpful in determining whether they want a child, or they feel they should have a child.

> After carefully exploring her motivations Victoria, age thirty-eight, came to the conclusion that she did want a child. 'Because my childhood had been so unhappy I believed history would repeat itself. I understand now that I don't have to recreate the past in the present. I have the financial resources, the maturity and the love to be a very good mother indeed.'

On the other hand, Elizabeth, age thirty-seven, discovered through therapy that her intense desire to have children was masking a deeper need to nurture herself. She was torturing herself about something she didn't really want. 'Whenever I used to hold a baby I would dissolve into tears. I felt so much love. I just thought this meant I wanted to have a baby of my own. In therapy I began to explore the emotion more fully and I began to understand that holding the baby was symbolically holding me, taking care of me, nurturing me. I realised that I needed to take better care of myself and that I didn't really want a child at all.'

Similarly Mary, age thirty-six, discovered that need for children was more about escaping the pressures of her job and having more fun in her life. 'With that insight the intensity of my desire for a child began to diminish. I took more time off work, started a new study programme and began to go out and travel abroad more. Within a year I knew that having my freedom was more important to me than having a child.'

Amy, age thirty-seven, found through exploring her emotions in therapy that her relationship with her mother was interfering with her decision about children. She was so frightened that she might be a bad mother because she herself had a difficult relationship with her mother. The issue of whether or not she could be a good mother was preventing her from grappling with the question of whether or not she even wanted to be a mother. 'Only after I confronted the issue of "am I good enough?" could I deal with the issue of choice.'

Support groups are another avenue for women to receive advice and support. It is powerful when women come together to explore issues, despite the many differences between them: some will be married, others are single, or living with partners, male and female. In support groups women can talk about all their conflicting feelings and receive understanding from others who share them. Hearing other women talk about their own issues can give you perspective on your own life and can help clarify your own decision-making process.

As the option of not having or delaying babies becomes more popular, support groups for women considering the baby-or-not option are likely to become more commonplace. In the USA women suffering from biological-clock anxiety can find local support groups by contacting RESOLVE (see resources section), an organisation for women contemplating motherhood. Courses like these tend to be more popular for women in their late thirties who feel their options are running out. In the UK a woman can get advice from her local GP, or she may prefer to contact a qualified therapist and counsellor, or some of the advisory groups listed in the resources section of this book.

In the mid 1990s Michelle McGrath ran a series of three-day workshops in London entitled 'Baby or Not', but until such support groups become more commonplace in the UK, one-to-one counselling and therapy may be the only option. The essence of one-to-one counselling and therapy is similar to support group therapy. It is all about finding out what you really want.

The idea of having a baby can often override whether or not a woman actually wants a baby and can be a good parent. The purpose of therapy, support groups, counselling or workshops for women considering motherhood is not for you to resolve all emotional conflicts so that you can have children or decide to remain childless. It is to help bring intellectual clarity to what seems to be a totally irrational process.

Counselling and support groups help you when you are battling with the many issues involved to make a choice. The goal is to find out what you really want. They also help you keep a sense of perspective and remind you that, as important as the choice for motherhood is, it should not be thought of as the one choice which determines the quality of your life. Fulfilment as a mother, or as a child-free woman, depends on all the other actions and decisions of a lifetime.

Much of the therapy will centre around your fears: fear of missing out if you don't have children, fear of life not being good again if you do. Hopefully you learn that there is no overarching right answer, only a personal solution that is right for you. Hopefully, you will be able, in the words of Joan Jenkins, from Women's Health Concern, to 'stop all that worrying'. Jenkins points out that whatever choice you make there is no right or wrong answer. 'Having children is a wonderful experience, but being child-free is not the end of the world.'

Therapy for biological-clock anxiety can help you make up your mind. But therapy and counselling are not for everyone. You may feel more comfortable dealing with the issues yourself, or talking to partners, friends and family. However you choose to deal with the issue, the important thing is that you don't ignore it – that, without letting it take over your life, you do try to come to some kind of resolution that works for you.

## Making up your mind

In your twenties and early thirties you have more time to ponder the question of whether or not to become a parent. However, once a woman passes the age of thirty-five or thirty-six, and fertility starts to decline, as difficult as the choice of motherhood may be, it should be confronted head on. It is preferable to reach an enlightened decision by grappling with all the issues involved than to risk finding yourself childless or panicked into pregnancy by the biological clock.

There are tools and techniques recommended by experts to help undecided women cope with biological-clock anxiety. For instance, talking to women who have chosen not to mother or become mothers can be very helpful by bringing the key issues into focus and by offering insights into the consequences of living with each choice. Spending time with children to 'see what you may be getting yourself into' can be useful. Imagining your life with and without children can also be an interesting exercise.

But, although helpful, none of these techniques can really answer the most important question a woman must ask herself before she makes the choice: how much do I really want a child? If you are nearing readiness to become a parent, here is a brief review of the kind of questions you should be asking yourself to make sure a child is what you really want:

- Are you ready for the possible discomforts of pregnancy? Are you ready to endure the pain of childbirth?
- Are you willing to open up your relationship, if you have one, to a third party? Are you ready for your relationship to change?
- Are you willing to devote the first months of your baby's life entirely to him or her? Are you prepared to feel lonely and isolated at times?
- If you plan to raise a baby alone, are you ready to endure criticism and judgment? Are you willing to accept that it may be harder for you to find a partner?
- Are you prepared to get by with less sleep? To child-proof your house? To feel overwhelmed with the responsibility of a child? Are you willing to endure times of chaos? Are you willing to forgo times of personal pleasure? Are you prepared to give up most of your free time?
- Have you thought about the difficult times as well as the good times? What if the baby has a health defect? What if your partner dies? What if you lose your job?
- Can you consider every stage of parenthood, not just the baby years, but the toddler and the teenage years?
- Are you prepared to love your child whatever the sex, personality or temperament?
- Are you prepared to relocate for the sake of your child?
- Are you prepared for the cost and extra expenses a child will bring?
- Are you prepared to lose control over your life?
- Are you prepared for the fact that time is no longer going to be your own?
- If you plan to work, are you prepared for the stresses of juggling home and career life?
- Are you prepared for the physical and emotional changes of motherhood?

## Giving yourself permission to wait

During the decision-making process it may become obvious to you that you are not ready to become a mother at this time. You think you need more time to make a decision.

It is preferable to make a decision about motherhood before your mid to late thirties. Sitting on the fence is stressful. But many women find that their views constantly shift and they simply can't decide. If this is the case, waiting is advisable.

The waiting while you make up your mind and sort out your priorities may at times feel agonising, but comfort yourself with the thought that there are good reasons why you don't have a baby yet. Even if your biological clock is ticking so loudly you can't sleep at night, it is better to give yourself more time than to have a child because you think you ought to.

Waiting is especially advisable if you don't feel emotionally ready, if you are not prepared for the stresses and sacrifices of having a child, if your lifestyle or relationship is not conducive to raising children, and if you are not at all sure you want to make a lifetime commitment to a child.

Waiting to have a baby until you feel ready makes sense. Take comfort from the fact that more and more women are having babies in their late thirties and early forties. Coming to motherhood later does have its advantages. And whatever reason you have for waiting, see it as a positive step. Clear, realistic thinking to assess your potential as a parent will reassure you that when you do make up your mind the choice you make is right for you.

If you are keeping your options open, however, it is important that you do understand that you are taking a risk and that you are mature enough to handle the consequences. It is possible that the need to have a child could become overwhelming in your forties and fifties when you can't conceive and don't have a choice any more.

If this is the case, regret and sadness is understandable, but again remember there were reasons for the choices you made. And there are other ways to bring a child into your life, such as adoption or fostering. It's what happens to children after they are born that really matters.

## Allowing yourself the right to say no

In 1983 Wellesley College psychologists Grace Baruch and Rosalind Barnett in collaboration with Boston University journalism professor Caryl Rivers published a landmark study of women, family and work. It is called the *Lifeprints* study and it attempted to find out what women really want. The emotion-laden issue of whether or not to have children which the study documents is as relevant today as it was in the early 1980s.

Women in the *Lifeprints* study tended to think in extremes: having children means loss of freedom; not having them means feeling empty and unfulfilled. But the study also revealed that the question of whether or not to mother is not straightforward, but influenced by individual life circumstances, relationships, goals and aspirations. It also shows that motherhood simply does not suit a number of women. The simple idea that women must mother to feel completely fulfilled is refuted.

According to the *Lifeprints* study, and other more recent surveys, not having children does not mean that a woman feels empty and functionless, unless she wants children and can't come to terms with not having them. In fact, women come across as far more hesitant about children and far more realistic about the frustrations and sacrifices they involve than men.

Even though a child is inherent to the traditional idea of a woman, many women today are choosing other ways to mother. A woman can mother other people, other people's children, siblings, parents, ideas, communities, projects, jobs, plants, animals and so on. Children can still be a part of her life if she chooses. She can make an incredible impact on their lives in other ways by teaching, mentoring, role-modelling. There are opportunities to befriend and tutor a child, become an honorary parent, sponsor a child from another country, work with children in a professional or voluntary capacity and so on. There are so many contributions to make to a child's development. It is high time that we began to value these contributions as highly as the birthing experience.

You may well find that motherhood is not for you. Perhaps you have other priorities in life, or you feel that your lifestyle or temperament wouldn't suit a child. Whatever the reason, every woman has a right to choose the lifestyle that suits her.

Making a decision can be a great relief, but remember, there will be times when you will feel regret too. Deciding not to have children involves coping with the inevitable fear of missing out later in life. Keeping in touch with the reality of parenthood instead of the fantasy, and reminding yourself of all that you have achieved in your life because of your decision not to have children, can be helpful. Think too of all the energy, time, freedom and extra money you have. Remember also that mothers have their regrets and frustrations too. No life choice is ever perfect.

There is no reason a woman without children should feel alone. The number of women choosing to look at what they have to offer other than motherhood is increasing. According to the US census-takers, nearly 20 per cent of women in the baby-boomer generation are not having children. The US has a network, called ChildFree, with over fifty meeting places for child-free people. The organisation believes that child-free people should be respected for who they are and not be judged by whether or not they have children. A child-free life can be full, productive and happy and should be carefully considered by all those not sure about children.

In the UK, ISSUE is currently in the process of setting up an organisation for child-free people called 'MoreToLife'. The aim of this exciting new development is to have a network of support across the UK for the growing number of child-free people. If you have chosen not to have children, it might be worth contacting them or forming your own organisation for like-minded women and couples.

## A new beginning

We are a lucky generation of women. We may not 'have it all' yet, but we have a lot. We don't have to be dictated to by our biology any more. We can lead full and interesting lives with or without children. We have choice.

But the trouble with choice is that when you do finally opt for one thing other options disappear. There can be no right or wrong answer when a woman is struggling with the motherhood decision. It is a matter of each woman's individual life-script. What is right for you. But one thing is for certain, whatever choice you make there will be losses. Nothing in life offers perfect fulfilment. Have children and you miss out on freedom, personal space and spontaneity. Don't have children and you miss out on the joy and wonder a child can bring to your life. Decide to postpone the decision until you feel ready and you risk problems with fertility. Have children too soon and risk frustration and regret.

Fortunately, though, if you can accept and acknowledge the reality of loss, you will also discover that with every loss there also comes an opportunity for gain. When you understand why you haven't got children you are likely to appreciate how fulfilling freedom and other things in life are to you. And if you have children and can accept that being a mother involves a loss of freedom, you can begin to focus more on how you can positively recreate your life around your children.

Whether or not you decide to have children, there are opportunities for growth and fulfilment. Deciding about motherhood is a crisis point in your life. You discover what your priorities and possibilities are. You set the course for the rest of your life.

With so much female identity resting on motherhood, it can be hard if you are suffering from biological-clock anxiety to maintain a sense of optimism and perspective, but if the crisis can jolt you into reflection and careful assessment of how much you want to be a mother, this can only be positive. By questioning what you really want from life, how important motherhood is to you, you can emerge with a stronger sense of your own identity.

Seen in this light, your choice to be, or not to be, a mother is an exciting voyage of self-discovery. It marks not a disappointing end, but a new beginning.

# Afterword

*You're beginning to feel older. You heard the first tick of your biological clock at age 30, when a glint of silver appeared in your hair. And the ticking grew louder at 35, when crow-feet made their debut. And it reached a crescendo at 40, when you wondered whether you were fertile enough to get pregnant.*

(Patricia Fisher (ed.), *Age Erasers: Actions You Can Take Right Now to Look Younger and Feel Great*, *Prevention* Magazine Books (Rodale Press, Pennsylvania, 1994), p. 55)

Biological-clock anxiety often goes hand in hand with a fear of ageing. Knowing you are still fertile can make you feel young. Having a child is a way of delaying the ageing process.

But age happens. Sometime after the age of thirty when human growth hormones, oestrogen, progesterone and testosterone are reduced, we shrink rather than grow. Our skins begin to thin and become dry. The immune system isn't so efficient. Egg quality starts to decline and eventually we can't have babies any more.

Scientists are working on various strategies to delay the ageing process. Some are studying genes to correct genetic mistakes. Others are focusing on hormones that affect cell growth. Some are developing compounds that will trap free radicals, the molecules that age us. And some are studying how lifestyle changes can help. Fertility experts are finding ways for women to defy the biological clock. Strategies are being developed. Some experts believe that eating less can keep us healthier and more fertile, or that taking antioxidants can delay ageing, or that hormonal changes can make us look and feel younger. Remarkable developments in reproductive technology are giving women a fighting chance against the biological clock.

But medical science is still a long way from being able to stop the ageing process or guarantee fertility. Until scientists can find a way to turn back the clock, at some point every one of us has to come to terms with the fact that, along with less flexible joints, less acute vision and less trustworthy memory, our fertility will decline and then fade away.

Ageing and the loss of fertility can be unnerving even for the most liberated of women. In our hearts and minds we still feel like children, but in our mirrors we see something different. We fear that menopause heralds the beginning of the end – not just of fertility, but of sexuality, desirability, and relevance in today's society.

But this isn't the case at all. When reproductive function ends a new and exciting chapter begins.

Once again it's a matter of accepting realities at each stage of life. Of adjusting to the notion that you are getting older. Of seeing not just the negatives but the positives. It is difficult giving up the miracle of fertility, but when you do, like any other loss that you have the courage to accept and acknowledge, what you have to lose is not nearly so valuable as what you have to gain. You give yourself the chance to replace what has been sacrificed with something better.

The menopause is often described as the 'change of life', but, whatever you call it, no other time in a woman's life has as much potential for creativity, fulfilment, excitement and understanding the true source of your power as this one. 'On the other side of all that turmoil, there is the most wonderful moment in one's whole life – really the most golden, the most extraordinary, luminous instant that will last forever' (Germaine Greer, quoted in Melinda Beck, 'Menopause', *Newsweek*, 25 May 1992, p. 38).

Many women say that they feel better about themselves at this stage in their lives than at any other time. Ironic isn't it! Only when your biological clock stops ticking at menopause do you finally realise that time may, in fact, have been on your side all along.

# Useful Addresses

## Advice, information and support for women contemplating motherhood or the child-free lifestyle

American Association for Marriage and Family Therapy
1133 15th Street, NW, Suite 300
Washington, DC 20005
USA
Telephone: 202 452 0109
Website: www.aamft.org

ChildFree
7777 Sunrise Boulevard, Suite 1800
Citrus Heights, CA 95601
USA
Telephone: 916 773 7178

National Women's Health Network
514 10th Street, NW, Suite 400
Washington, DC 20004
USA
Telephone: 202 628 7814
Website: www.womenshealthnetwork.org

RESOLVE: The National Infertility Association
1310 Broadway
Somerville, MA 02144
USA
Telephone: 617 623 0744
Website: www.resolve.org

Single Mothers by Choice
PO Box 1642
Gracie Square Station
New York, NY 10028
USA
Telephone: 212 988 0993
Website: www.parentsplace.com/family/singleparent

British Association for Counselling and Psychotherapy
1 Regent Place
Rugby
Warwickshire CV21 2PJ
UK
Telephone: 0870 443 5252

Family Planning Association
2–12 Pentonville Road
London N1 9FP
UK
Telephone: 020 7837 5432

ISSUE: The National Fertility Association
114 Lichfield Street
Walsall WS1 1SZ
UK
Telephone: 01922 722888

Maternity Alliance
45 Beech Street
London EC2P 2LX
UK
Telephone: 020 7588 8583

MoreToLife Initiative (A child-free network)
114 Lichfield Street
Walsall WS1 1SZ
UK
Telephone: 01922 722888
Website: www.moretolife.co.uk

## Fertility awareness

American College of Obstetricians and Gynecologists
409 12th Street, SW
Washington, DC 20090
USA
Telephone: 202 638 5577
Website: www.acog.org

American Infertility Association
666 Fifth Avenue, Suite 278
New York, NY 10103
USA
Telephone: 718 611 5083
Website: www.americaninfertility.org

American Menopause Foundation
350 Fifth Avenue, Suite 2822
New York, NY 10118
USA
Telephone: 212 714 2398
Website: www.americanmenopause.org

American Society for Reproductive Medicine (formerly the American Fertility Society)
1209 Montgomery Highway
Birmingham, AL 35216-2809
USA
Telephone: 205 978 5000
Website: www.asrm.org

Society for Assisted Reproductive Technologies (SART)
1209 Montgomery Highway
Birmingham, AL 35216-2809
USA
Telephone: 205 978 5000
Website: www.sart.org

Family Planning Association
2–12 Pentonville Road
London N1 9FP
UK
Telephone: 020 7837 5432

FORESIGHT: Association for the Promotion of Pre-Conceptual Care
28 The Paddock
Godalming GU7 1XD
UK
Telephone: 01483 427839

Human Fertilisation and Embryology Authority (HFEA)
Paxton House
30 Artillery Lane
London E1 7LS
UK
Telephone: 020 7377 5077
Website: www.hfea.gov.uk

ISSUE: The National Fertility Association
114 Lichfield Street
Walsall WS1 1SZ
UK
Telephone: 01922 722888

## Women's health

Office on Women's Health (Department of Health and Human Services)
200 Independence Avenue, SW, Room 730B
Washington, DC 20201
USA
Telephone: 202 690 7650
Website: www.4woman.gov/owh

Women's Health Information Center (Journal of the American Medical Association)
USA
Website: www.ama-assn.org/special/womh

Women's Health
52 Featherstone Street
London EC1Y 8RT
UK
Telephone: 020 7251 6580

## Adoption, fostering, mentoring, surrogacy

Big Brothers Big Sisters of America
230 North 13th Street
Philadelphia, PA 19107
USA
Telephone: 215 567 7000
Website: www.bbbsa.org

National Council for Adoption
1930 17th Street, NW
Washington, DC 20009
USA
Telephone: 202 328 1200
Website: www.ncfa-usa.org

National Council for Single Adoptive Parents (formerly Committee for Single Adoptive Parents)
PO Box 55
Wharton, NJ 07885
USA
Website: www.adopting.org/ncsap

British Agencies for Adoption and Fostering (BAAF)
Skyline House
200 Union Street
London SE1 0LX
UK
Telephone: 020 7593 2000

COTS (Childlessness Overcome Through Surrogacy)
Loandhu Cottage
Gruids
Lairg
Sutherland IV27 4EF
UK
Telephone: 0906 680 0088

National Foster Care Association (NFCA)
87 Blackfriars Road
London SE1 8HA
UK
Telephone: 020 7620 6400

## Alternative therapies for infertility

American Association of Naturopathic Physicians
8201 Greensboro Drive, Suite 300
McLean, VA 22102
USA
Telephone: 703 610 9037
Website: www.naturopathic.org

American Holistic Medical Association
6728 McLean Village Drive
McLean, VA 22101
USA
Website: www.holisticmedicine.org

Institute for Complementary Medicine
PO Box 194
London SE16 7QZ
UK
Telephone: 020 7237 5165

Institute for Optimum Nutrition
Blades Court
Deodar Road
London SW15 2NV
UK
Telephone: 020 8877 9993

# Bibliography


Armstrong, Pamela, *Beating the Biological Clock: The Joys and Challenges of Late Motherhood*, Headline, London, 1996

Bartlett, Jane, *Will You Be Mother?: Women Who Choose to Say No*, Virago, London, 1994; New York University Press, New York, 1994

de Beauvoir, Simone, *The Second Sex*, Knopf, New York, 1953

Belsky, Jay and John Kelly, *The Transition to Parenthood: How a First Child Changes a Marriage*, Delacorte Press, New York, 1994

Bombeck, Erma, *Just Wait Till You Have Children of Your Own*, Ballantine Books, New York, 1972

Bombeck, Erma, *Motherhood: The Second Oldest Profession*, Dell Publishing, New York, 1983

Bing, Elizabeth and Libby Colman, *Having a Baby After Thirty*, Bantam, New York, 1980

Borysenko, Joan, *A Woman's Book of Life: The Biology, Psychology and Spirituality of the Feminine Life Cycle*, Riverhead, New York, 1996

Bristow, Wendy, *Single and Loving It*, Thorsons, London, 2000

Carter, Jean and Michael Carter, *Sweet Grapes: How to Stop Being Infertile and Start Living Again*, Perspective Press, Indianapolis, 1989

Cheung, Theresa, *Androgen Disorders in Women: The Most Neglected Hormone Problem*, Hunter House, California, 1999

Cheung, Theresa, *The Thirties: A Woman's Guide to Health in the Decade of Transition*, Adams Media, Massachusetts, 2001

Clements, Marcelle, *The Improvised Woman: Single Women Reinventing Single Life*, W. W. Norton, New York, 1998

Cowan, Carolyn and Philip Cowan, *When Partners Become Parents: The Big Life Change for Couples*, Basic Books, New York, 1992

Crittendon, Danielle, *What Our Mothers Didn't Tell Us: Why Happiness Eludes the Modern Woman*, Simon & Schuster, New York, 1999

Curtis, Glade, *Your Pregnancy After 30*, Fisher Books, Arizona, 1996

Dinnerstein, Dorothy, *The Rocking of the Cradle and the Ruling of the World*, Souvenir Press, London, 1978

Dockett, Lauren and Kristen Beck, *Facing 30: Women Talk about Constructing a Real Life and Other Scary Rites of Passage*, New Harbinger, Oakland, CA, 1998

Domar, Alice and Henry Dreher, *Healing Mind, Healthy Woman*, Henry Holt, New York, 1996

Eisenberg, Arlene, Heidi Eisenberg Murkoff and Sandee Eisenberg Hathaway, *What to Expect When You're Expecting*, Workman, New York, 1991

Emecheta, Buchi, *The Joys of Motherhood*, Heinemann International, 1994

Engel, Beverly, *The Parenthood Decision: Discovering Whether You are Ready and Willing to Become a Parent*, Doubleday, New York, 1998

Eyer, Diane, *Mother–Infant Bonding: A Scientific Fiction*, Yale University Press, New Haven, CT and London, 1992

Fabe, Marilyn and Norma Wikler, *Up Against the Clock: Career Women Speak on the Choice to Have Children*, Warner, New York, 1980

Fenlon, Arlene, *Getting Ready for Childbirth: A Guide for Expectant Parents*, Little, Brown and Co., Boston, 1986

Figes, Kate, *Life After Birth: What Even Your Friends Won't Tell You about Motherhood*, Viking, London, 2000

Fleming, Ann, *Motherhood Deferred: A Woman's Journey*, Fawcett Columbine, New York, 1994

Franklin, Robert and Dorothy Brockman, *In Pursuit of Fertility: A Fertility Expert Tells You How to Get Pregnant*, Henry Holt, New York, 1995

Friday, Nancy, *My Mother, Myself*, Fontana, London, 1988

Friedan, Betty, *The Feminine Mystique*, W. W. Norton, New York, 1963; Penguin, London, 1992

Friedman, Ann, *Work Matters: Women Talk About Their Jobs and Their Lives*, W. W. Norton, London, 1996

Furse, Anna, *The Infertility Companion: A User's Guide to Tests, Technology and Therapies*, Thorsons, London, 1997

Gavron, Hannah, *The Captive Wife: The Conflicts of Housebound Mothers*, Routledge, New York, 1984 (out of print)

Gerson, Kathleen, *Hard Choices: How Women Decide about Work, Career and Motherhood*, University of California Press, London, 1985

Gilman, Lois, *The Adoption Resource Book*, HarperCollins, New York, 1992

Gittleman, Louise, *Before the Change: Taking Charge of Your Perimenopause*, HarperSanFrancisco, 1998

Greer, Germaine, *The Whole Woman*, Doubleday, London, 1999

Hales, Dianne, *Just Like a Woman: How Gender Science is Redefining What Makes Us Female*, Virago, London, 1999

Hays, Sharon, *The Cultural Contradictions of Motherhood*, Yale University Press, London, 1996

Hickey, Mary and Sandra Salmans, *The Working Mother's Guilt Guide: Whatever You're Doing, It Isn't Enough*, Penguin Books, New York, 1992

Hite, Shere, *The Hite Report*, Pandora Press, London, 1989

Iovine, Vicki, *The Girlfriend's Guide to Pregnancy*, Perigee, New York, 1996

Ireland, Mardy, *Reconceiving Women: Separating Motherhood from Female Identity*, Guildford Press, New York, 1993

Jeffers, Susan, *I'm Okay . . . You're a Brat*, Hodder & Stoughton, London, 1999

Jones, Cheryl and Bruce Rappaport, *The Adoption Sourcebook: A Complete Guide to the Complex Legal, Financial and Emotional Maze of Adoption*, Lowell House, California, 1999

Jones, Maggie, *Motherhood After 35: Choices, Decisions, Options*, Fisher Books, Arizona, 1998

Klein, Carol, *The Single Parent Experience*, Avon, New York, 1973

Lafayette, Leslie, *Why Don't You Have Kids? Living a Full Life Without Parenthood*, Kensington Books, New York, 1995

Lerner, Harriet, *The Mother Dance: How Children Change Your Life*, HarperCollins, New York, 1998

Levison, Daniel, *The Seasons of a Woman's Life: A Fascinating Exploration of the Events, Thoughts, and Life Experiences That All Women Share*, Alfred A. Knopf, New York, 1996

McConville, Brigid, *Mad to Be a Mother: Is There Life after Birth for Women Today?*, Century, London, 1987

McKaughan, Molly, *The Biological Clock*, Penguin, New York, 1989

McKenna, Elizabeth Perle, *When Work Doesn't Work Anymore: Women, Work and Identity*, Simon & Schuster, London, 1997

Maitland, Sara, *Why Children?*, ed. Dowrick and Grundberg, Women's Press, London, 1980

Mandarin, Hope (ed.), *The Handbook for Single Adoptive Parents*, Committee for Single Adoptive Parents, PO Box 15084, Chevy Chase, MD 20825, 1992

Mattes, Jane, *Single Mothers by Choice: A Guidebook for Women Who Are Considering or Have Chosen Motherhood*, Times Books, New York, 1994

Matus, Irwin and Arwin Matus, *Wrestling with Parenthood: Contemporary Dilemmas*, Gylantic Publishing Co, New York, 1995

Maushart, Susan, *The Mask of Motherhood: How Mothering Changes Everything and Why We Pretend It Doesn't*, Random House, Australia Pty, Ltd, 1997

Miller, Naomi, *Single Parents by Choice: A Growing Trend in Family Life*, Insight Books, New York, 1992

Morris, Monica, *Last-Chance Children*, Columbia University Press, New York, 1988

Northrup, Christiane, *Women's Bodies, Women's Wisdom*, Bantam, New York, 1998

Paulson, Richard and Judith Sachs, *Rewinding Your Biological Clock: Motherhood Late in Life*, W. H. Freeman, New York, 1998

Prevention Biological Clock Anxiety Website: http://www.healthyideas.com/healing/cond_ail/bioclock.html

Robinson, Susan, *Having a Baby Without a Man: The Woman's Guide to Alternative Insemination*, Simon & Schuster, New York, 1985

Safer, Jean, *Beyond Motherhood: Choosing a Life Without Children*, Pocket Books, New York, 1996

Salzer, Linda, *Surviving Infertility: A Complete Guide through the Emotional Crisis of Infertility*, HarperPerennial, New York, 1997

Schwartz, Judith, *The Mother Puzzle: A New Generation Reckons with Motherhood*, Simon & Schuster, New York, 1993

Sheehy, Gail, *The Silent Passage*, Random House, New York, 1991

Sheehy, Gail, *New Passages: Mapping Your Life Across Time*, HarperCollins, London, 1996

Siegal, Paula, *The Female Body: An Owner's Manual*, *Prevention* Magazine Health Books, Rodale Press, Emmaus, Pennsylvania, 1996

Silber, Sherman, *How to Get Pregnant With the New Technology*, Warner Books, New York, 1991

Spillane, Mary and Victoria McKee, *Ultra Age: Every Women's Guide to Facing the Future*, Macmillan, London, 1999

Stoppard, Miriam, *Woman's Body: A Manual for Life*, Dorling Kindersley, London, 1994

Swigart, Jane, *The Myth of the Bad Mother: The Emotional Realities of Mothering*, Doubleday, New York, 1991

Thorn, Gill, *Not Too Late: Having A Baby After 35*, Bantam, London, 1998

Tilsner, Julie, *29 and Counting: A Chick's Guide to Turning 30*, Contemporary Books, Illinois, 1998

Tobin, Phyllis, *Motherhood Optional: A Psychological Journey*, Jason Aronson, London, 1998

Vercollone, Carol Frost and Robert and Heidi Moss, *Helping the Stork: The Choices and Challenges of Donor Insemination*, Macmillan, New York, 1997

Vliet, Elizabeth, *Screaming to Be Heard: Hormonal Connections Women Suspect and Doctors Ignore*, M. Evans, New York, 1995

Wagner, Laurie, *Expectations: Thirty Women Talk about Becoming a Mother*, Chronicle Books, San Francisco, 1998

Walter, Carolyn, *The Timing of Motherhood: Is Later Better?*, Lexington Books, London, 1986

Welldon, Estella, *Mother, Madonna, Whore: The Idealization and Denigration of Motherhood*, Free Association Books, London, 1988

Weschler, Toni, *Taking Charge of Your Fertility*, HarperPerennial, New York, 1995

Weston, Carol, *From Here to Maternity: Confessions of a First-Time Mother*, Little, Brown and Co., London, 1991

Winston, Robert, *Getting Pregnant: The Complete Guide to Fertility and Infertility*, Pan, London, 1993

Wiscot, Arthur and David Meldrum, *Conceptions and Misconceptions: A Guide Through the Maze of In-Vitro Fertilization and Other Reproductive Technologies*, Hartley & Marx, Point Roberts, WA, 1997

Woitiz, Janet G., *Healthy Parenting: An Empowering Guide for Adult Children*, Simon & Schuster, New York, 1992

# Index